ROUTLEDGE LIBRARY EDITIONS:
JAPAN

SHINGŪ

SHINGŪ

A Japanese fishing community

ARNE KALLAND

Volume 75

Routledge
Taylor & Francis Group
LONDON AND NEW YORK

First published in 1981

This edition first published in 2011
by Routledge
4 Park Square, Milton Park, Abingdon, Oxon OX14 4RN
605 Third Avenue, New York, NY 10017

Routledge is an imprint of the Taylor & Francis Group, an informa business

British Library Cataloguing in Publication Data
A catalogue record for this book is available from the British Library

ISBN 13: 978-0-415-56498-4 (Set)
eISBN 13: 978-0-203-84317-8 (Set)
ISBN 13: 978-0-415-58805-8 (hbk) (Volume 75)
eISBN 13: 978-0-203-84398-7 (Volume 75)

Ill.1 A retired fisherman on the way home from the harbour with fish for supper.

SCANDINAVIAN INSTITUTE OF ASIAN STUDIES
MONOGRAPH SERIES NO 44

Shingū

A Japanese fishing community

Arne Kalland

Curzon Press

Scandinavian Institute of Asian Studies
Kejsergade 2, DK-1155 Copenhagen K

First published 1981
Curzon Press Ltd: London and Malmö

ISBN 0 7007 0136 2
ISSN 0069 1712

Publication of this book was aided by a grant from the Norwegian Research Council for Science and the Humanities

Printed in Great Britain by
Biddles Ltd, Guildford, Surrey

CONTENTS

LIST OF TABLES

PAGE

LIST OF MAPS

LIST OF ILLUSTRATIONS

PAGE

LIST OF FIGURES

PREFACE

The field work on which this book is based was financed by a scholarship given by the Japanese Government, Ministry of Education. Thanks to a grant from the Scandinavian Institute of Asian Studies I was able to re-visit the field in May-June 1978. The Norwegian Research Council for Science and the Humanities has contributed funds for the publication of this book.

I have received valuable help from many persons and institutions. At Kyūshū University where I was affiliated, I thank Prof. Nakamura Masao who helped me to get into Shingū, the fishing community in which my wife and I lived for almost a year from May 1975. I also thank Prof. Naito Kanji and Kuroda Yasuo.

The library at Fukuoka Cultural Center was most helpful in interpreting historical documents, and I extend my thanks particularly to Miss Kitamura Keiko and Mr. Hirowatari Masanori.

During our first month in Kyūshū when we were looking for a suitable community to study, we lived at the Norwegian Seamen's Mission in Moji, Kitakyūshū City, and I take the opportunity to thank the manager Kjell Hodne and his lovely wife Else for their kindness both in the difficult initial period, and later when we visited the mission.

I am also indebted to friends and colleagues at the Institute for Social Anthropology, to Fredrik Barth at the Ethnographic Museum, and to Inadomi Masahiko and Yamaguchi Takafumi at the East Asian Institute, all in Oslo. At the Scandinavian Institute of Asian Studies in Copenhagen I have received invaluable comments from Carl T. Steenstrup.

But of course, my deepest feeling of gratitude goes to the people in Shingū; first of all to the fishermen who never expressed irritation over my ignorance, but also to the people at the town office and the fishing cooperative, the Shintō priest Shinozaki Asao, and the Buddhist priest Shibata Toshiki. And special thanks to Mr. and Mrs. Shinozaki Kazuichi, the manager of the cooperative, who opened their doors for us, and made their home ours for ten unforgettable months, and again in May and June 1978.

Finally, I wish to thank my wife Momoyo M. Kalland who helped me both in Shingū and later. This book would never have been written without her help.

Shingū, March 1980. Arne Kalland

PROLOGUE

The following short essay was written by a fisherman's daughter as homework for the school where she attended the sixth grade. Her father is one of the most industrious fishermen in Shingū, and he became one of our very best friends.

MY FATHER
by Kirishima Kaori

My father is forty-one years old and he is a very healthy father. Father is the centre of the family and helps everybody.

A long time ago, they say, my father was a head clerk in a shop somewhere. He worked there for some years, and then he started to work at Hakata city-office. After that, I heard, he became a fisherman in order to take over the house. My father is still working as a fisherman. I am very glad that my father is a fisherman.

My father's hands are always working with things like thick ropes; his hands are hard and thick, his fingers thick. When it's really bad the fingernails sometimes come off. His face is dark the whole year round so I have never seen my father's skin peeling off during the summer.

Father drinks *sake* (rice wine) very often.

I don't know for sure why my father became a fisherman, but I know one reason. That is because father is the first son of this house, so I think he became a fisherman in order to take over the house. But I don't think he became a fisherman only for that reason. I think I'll ask my father another time.

Whenever he speaks of his past, he always tells me that he had a good head and was good at sport when he went to elementary and middle school.

My father, who is kinder than my mother, used to buy me picture-books when I was a child. But when such a kind father gets angry, I get very scared.

If my brother and I quarrel, he often says: "Why do you quarrel? If you quarrel that much, get out of here both of you!"

Or if I make a mess on the desk, he hits my head and says: "Why do you make it so untidy? And you're a woman!" Father always lets me do anything because, he says, I am a woman. Sometimes I dislike father behaving in this way. Sometimes I think it's strange that I both like and dislike my father.

My father's cooking speciality is roast meat. He is better at frying meat than mother. Father often plays games and *sumō* (Japanese wrestling) with my brother and me, but I strongly dislike father coming home after drinking *sake*. Because the smell of *sake* is strong and makes me feel bad, and because he then quarrels with my mother. I also sometimes drink beer offered me by my father, but I don't like *sake.* Then my mother says: "You should never marry a drunkard". I think that if he didn't drink *sake,* he would be a very good father.

CHAPTER I

INTRODUCTION

Japan's fisheries can be divided into three categories: (1) Pelagic fisheries which operate in far-away waters all over the world. (2) Offshore fisheries which are conducted by fishing boats being on the fishing grounds outside the archipelago for about a week at a time. (3) Coastal fisheries which are operated by small craft leaving and returning to their home bases every day. Though the coastal fisheries only amount to a fraction of the total catches, they engage the majority of the fishing enterprises.[1] The majority of the fishermen are engaged on these small craft which have their bases in small fishing communities spread all over the Japanese coast. This book will discuss one of these communities, namely Shingū.

Although Japan, with her annual catches of more than ten million tons, is the largest fishing nation in the world, little is known about the sociology of the fisheries. Borgstrom (1964) treats the Japanese fisheries in a global perspective and has little space to discuss the men behind the catches. The same is true regarding a large range of essays about specific Japanese fisheries, usually of the offshore or pelagic kind.

A few notable exceptions exist in English. The first that appeared in 1954 is an extensive ethnographic account written by Norbeck. Takashima, as the village is called, is located in the Inland Sea well protected by islands. The inhabitants were originally farmers who turned to fishing only after the Meiji Restoration in 1868, and they were still part-time farmers in 1954. The second was written by Brameld (1968) and discusses a stigmatized fishing community which he called Ebibara. The book focuses on education and values in Ebibara and compares these with those of the *burakumin* outcaste and the majority populace. The last book to be mentioned here is written by Bernier (1975) and focuses on the religious life in a village called Sone situated in a bay in Mie prefecture.

The object of this book is partly to enrich the ethnographic knowledge on the Japanese fishing population, and thus contribute to the literature on what Casteel and Quimby call "maritime anthropology" (Casteel and Quimby 1975).

Shingū is situated in other surroundings than Takashima, Ebibara and Sone. Shingū faces the rough Japan Sea which has implications both for the climatic conditions as well as for the species caught. Furthermore, it has a long history as a fishing village (cf. Chapter III) and is not stigmatized as Ebibara. Whether Shingū is more "representative" than the others is on the other hand an open question. Historical documents indicate that Shingū might have been fairly representative in its way of fishing for the Northern Kyūshū area, and similar traits have been reported from the Kujūkurihama district in Chiba prefecture outside Tōkyō (Kujukuri Research Committee 1958). Moreover, the main fish caught by Shingū fishermen during this early period was sardine, and more than half the quantity taken around Japan before World War II was sardine (Moyer 1965:136n). This had far-reaching consequences for the social structure of the fishing villages because the sardine fisheries required big nets like the purse seine or beach seine, which led to large management units (or net-groups) that are absent in Takashima, Ebibara and Sone. These management units were often owned by capitalists in the fishing villages, and this brings us to the concept of *amimoto* (lit. the origin of the net, i.e. net owner). As Takashima and Ebibara did not have net-groups, they did not have *amimoto* either.

This book will focus on these net-groups and other forms of management units found in Shingū, i.e. those based on independently operated small fishing boats. More specifically, I will focus on the structure and leadership of these units. To be a successful leader "is to gain access to more resources than one's opponents and to use them with greater skill" (Bailey 1969:36), and the leadership can thus be regarded as being generated from some basic factors within the village, namely access to and control of the following resources: (i) capital, (ii) labour, (iii) fishing rights, and (iv) markets.

Profound changes have occurred in Japan since the industrialization started after the Meiji Restoration. These changes have led to: (1) Increased economic opportunities and a wider range of objects available for investments. (2) Increased physical and social mobility, and Shingū's proximity to big industrial centres is important here. (3) Changes of the laws and administrative apparatus. I will argue that the changes (1) – (3) have affected seriously the access and control of the resources mentioned above, which in turn have led to changes in the leadership patterns both on the boats and in the net-groups. I will try to show how the resources i–iv could under certain conditions be monopolized in the hands of a merchant. As an *amimoto* he controlled by definition the capital and labour and frequently the market. The village headman had also a crucial position as had likewise the district headman over a number of fishing villages. The district headman had the decisive power over a fund made available from which the fishermen could borrow money for boats, nets and houses. Judicial authority was given to both the village and the district headmen, and both had influence over the fishing grounds. Where the *amimoto* and the headman were the same person, this person came to control all the resources mentioned, and he exercised a profound influence over the fishermen.

After the Meiji Restoration the situation changed on the lines given in (1) – (3) above. I will try to show that the merchants lost control over the resources. Consequently, their dominant position in the fisheries declined and gave way for a more egalitarian organization of the fisheries and to the establishment of fishing associations which got control over three of the resources, namely the capital, the market and the fishing rights. I will further argue that environmental changes have strengthened this tendency. The sardine which was the basis for the large net-groups with big and expensive nets, disappeared and is no longer hunted. The new species caught did not give rise to such large structures. The net-groups today mean less to the total income than previously, and the "functional importance" (Davis & Moore 1945) of the leaders has declined. Anybody is able to fill this position at present. Equal status in the net-groups has been institutionalized by the leadership being rotated among the members, or in some cases the leaders are elected by secret balloting. Such practices are old and well-known in Japan and are thus not attributable to diffusion of "modern democratic ideas" from the West. In fact, the rotational system has been used in Shingū at least since the beginning of the nineteenth century when the position as village headman seems to have changed in this way. It was natural that the fishermen looked to this kind of institution when they formed their new egalitarian net-groups in the 1930s.

Increased capitalization has in many places been accompanied by changes in patterns of ownership, and frequently a centralization of capital has occurred. Firth, in his revised edition of *Malay Fishermen*, states that when the British came to the Malacca peninsula "social and economic ties between the peasant and the feudal lord tended to disappear. While this meant greater freedom, security and individuality of the peasant as a whole, it has facilitated economic differentiation between classes of peasants" (Firth 1966:296). The same process he observed among the fishermen as the capitalization of the business progressed. Fukutake (1972:10) points out that the number of tenants increased in Japan since the Meiji Restoration, as did the number who had to rent some of the land they cultivated. A similar development has not occurred in Shingū, and we have to ask why increased differentiation did not take place among the Shingū fishermen.

I think there are several reasons for this. First, the big net-groups based on expensive nets disappeared. Unlike the situation described by Firth, the merchants withdrew from the purse seine in Shingū at a time when no fishermen had the capital needed to invest in these expensive nets. For lack of outside capital they were forced to unite. Secondly, much of the fishing was best done by small units which encouraged many small boats into operation. Thirdly, the fishing grounds became an estate for a corporation of fishermen. Any member of the fishing association was allowed to fish with line though netting usually required additional licences. Fishing rights could not be mortgaged and lost as could the land. Fourthly, the harbour facilities in Shingū were inadequate for harbouring large vessels, and thus Shingū did not attract capital for these.

Fifthly, and this point I think is the most important, the pattern of recruitment to the fishing industry led to increased homogeneity. I will argue that during the Tokugawa period (1603–1868) the fishermen were a heterogeneous group. A few owned big nets, many more owned small ones, others used only lines which required a minimum of capital investment, and some did not own any capital at all. There was a place for everybody partly because the small mean household size (cf. Chapter IV) during this period made it necessary to hire crew-men in order to man the boats. I will further argue that today there is an occupational mobility among the propertyless fishermen. The same process has been going on in many fishing villages, but contrary to many of them, this has not caused depopulation of Shingū, a fact that can be attributed to its proximity to Fukuoka City. The improved opportunities for work outside the fisheries have also caused a downwards mobility of the fishermen on the very top. Due to lack of crews they have been forced to reduce their scale of operation and join net-groups on the same terms as the majority. I will try to show how the homogeneity (and the threat of finding other employment) influences the structure of the leadership both in the net-groups and on the boats. The control of the labour force has been lost both to the *amimoto* and the skipper.

To pursue my aims I intend to use some historical data which are of two kinds: oral tradition and written documents. Very often I became aware of this historical information through tales told me by the informants, and later I frequently was able to verify their tales through documents unknown to these informants. Where this procedure has not been possible due to incomplete documents, the oral tradition has either been treated tentatively or has been omitted altogether. A further point should be mentioned: the documents found in Shingū do not give any complete picture of the community in the past, and I will thus supplement my data from Shingū with information from nearby fishing communities in similar ecological and societal circumstances.

My aims in using historical data are two. The first is to analyse and compare different management contexts in timeless settings. No fishing community shows all these forms at any time. A synchronic study could perhaps have been conducted if several communities with different technologies and economic structures were studied. In this way the shortcomings of the historical data could have been avoided. The method would on the other hand have required much more time or a team, and I had neither.

It would be clear from what I have mentioned so far, that fundamental changes have occurred within the fishing industry, and this brings me to my second object in using historical data, which is to analyse the development of the fisheries and their institutions. The shortcomings of my historical data are, I feel, balanced by my second object.

This book is based on several sources of information. The historical data have been discussed at some length, but something has to be said about their character. The following documents have been consulted in parts: (1) Akaishi Documents *(Akaishi monjo)* kept at the library of Fukuoka Cultural Center in Fukuoka City. Akaishi was the district headman for fourteen fishing villages during a brief period of the nineteenth century. The documents comprise mostly correspondence between Akaishi and the village headmen. A few of the documents have been published (Komonjo o yomukai 1976, Kaneuchi 1978). (2) Kaneuchi Documents *(Kaneuchi monjo)*. Kaneuchi was for a long time the leading house in Shingū, and was the district headman for the fourteen fishing villages before Akaishi. The family is now extinct, and the documents are scattered. Nonetheless, many have been traced and copies are kept at Fukuoka Cultural Center. The documents are in character similar to Akaishi's, and a few have been published (op.cit.). (3) Yamazaki Documents *(Yamazaki monjo)* are kept at Kyūshū University. Yamazaki was for a period village headman in Hakozaki fishing village (the same village as Akaishi's) and became the assistant of the district headman. Some of the documents are published in *Fukuoka-ken shi* (Fukuoka-ken 1963). These give valuable information on the administration and regulations in the fishing villages. (4) Isozaki shrine Documents *(Isozakigū monjo)* kept at the local shrine in Shingū. These documents mostly tell about the history of the shrine. (5) Yayama Documents *(Yayama monjo)* consist of five surviving documents telling about the Yayama family, a wealthy merchant house in Shingū during the Tokugawa period and the village headman and perhaps the district headman in the 1840s. Copies of the documents are kept at Fukuoka Cultural Center. (6) Books kept at the local temple. These books give some aid in tracing genealogies. (7) Some books kept by the hamlet headman give information about land ownership and residence, people holding hamlet positions, and events and regulations mostly around 1910.

Different kinds of sources were found at the town office (most of demographic character) and at the office of the fishing cooperative (mostly on income and quantities caught). By far the most important source of information was, however, the fishermen themselves. My wife and I lived in Shingū from May 1975 to March 1976 and again for a brief period during May–June 1978. Our home was unfortunately situated at some distance from the main cluster of fishing households (see Map 4), but I feel this was compensated for by our living with the family of the manager of the fishing cooperative. The information was obtained both by informal chatting and by formal guided interviews.

In December 1975 we conducted a community survey of some 250 households. (This was the total number of households in Shingū, the fishermen lived in 30 of these households.) The objects of this questionnaire were: (1) To check the validity of the impressions obtained through the daily contacts with the inhabitants, and to see whether these impressions were representative. (2) To get a basic idea about fundamental aspects of the community and its inhabitants, such as house ownership, religious affiliations, education etc. (3) To compare the fishing population with the rest of the community. Though little of the information obtained has been directly used, I feel nonetheless more confident in many of the statements that follow.

CHAPTER II

THE SETTING

The Term "Shingū"

Shingū-machi (or Shingū-chō; "machi" and "chō" are two alternative ways to read the Chinese ideogram meaning "township") is a result of several amalgamations of former independent villages. During the Tokugawa period the villages were to a great extent autonomous, based on one or a few hamlets. A distinction was made between farming villages *mura* and fishing villages *ura.* One fishing village was Shingū-ura. After the Meiji Restoration the distinction between these two kinds of villages was lifted, and our community became known as Shingū-mura. A sweeping reorganization of the villages was conducted in 1889, and five separate villages were merged to form the new Shingū-mura. These five were the farming villages of Shimonofu, Kaminofu and Minato plus the fishing villages of Shingū and Ainoshima. In 1954 the status of Shingū-mura was changed to township, and it became Shingū-machi. Finally, the next year Shingū-machi was merged with the neighbouring village of Tachibana-mura.

The old units from the Tokugawa period still exist, though they have no legal identity. The old *mura* and *ura* survive as *ōaza* (section), and this unit is again divided into *chiku, ku* or *buraku* (wards or hamlets). Ōaza Shingū consists of only one *chiku,* Shingū-ku.

The term "Shingū" is used by the inhabitants to refer to several different entities depending on the context. I will retain the term "Shingū", however, for the smallest unit, i.e. Shingū-ku, Ōaza Shingū and Shingū-ura which all cover the same geographical area. When referring to the larger administrative unit, "Shingū-machi" or "Shingū-mura" will be used.

The Location of Shingū

Shingū-machi which covers an area of 19.05 km^2, is located in Kasuya County of Fukuoka Prefecture on the northern shore of the island Kyūshū. Fukuoka, with its 4.3 million inhabitants, is one of the most densely populated prefectures in Japan. This population is not evenly distributed, however. The capital of the prefecture is Fukuoka City which is located just west of Shingū-machi. The city has just under one million inhabitants and is an amalgamation of the former castle city of Fukuoka and the port city of Hakata, but is only second in size to Kitakyūshū City which has slightly more than one million inhabitants. Kitakyūshū City is located on the northeastern fringe of the island facing the Kannon Strait separating Kyūshū from the main island Honshū, and the city is one of the most important industrial centres in Japan. Until the end of the last century the area was only a collection of small villages centred around the castle city Kokura, but grew spectacularly when the Yawata Steel Mill started operations in 1901. Kitakyūshū City is connected with Shimonoseki City on Honshū via railway, pedestrian and motor vehicle tunnels as well as a six-lane suspension bridge, one of the longest in the world. Kitakyūshū City is thus the gateway coming from Tōkyō or Ōsaka.

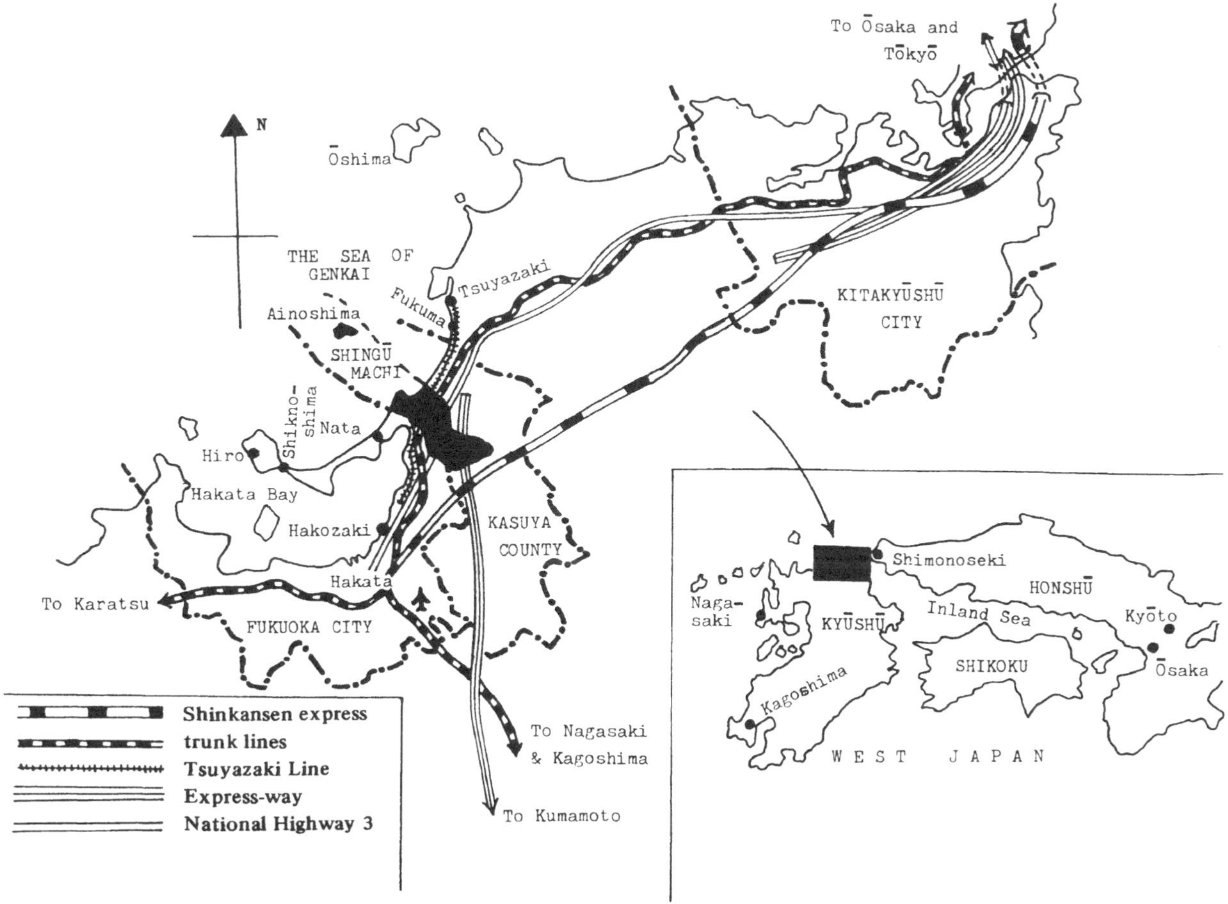

Map 1 Northern Kyūshū Region.

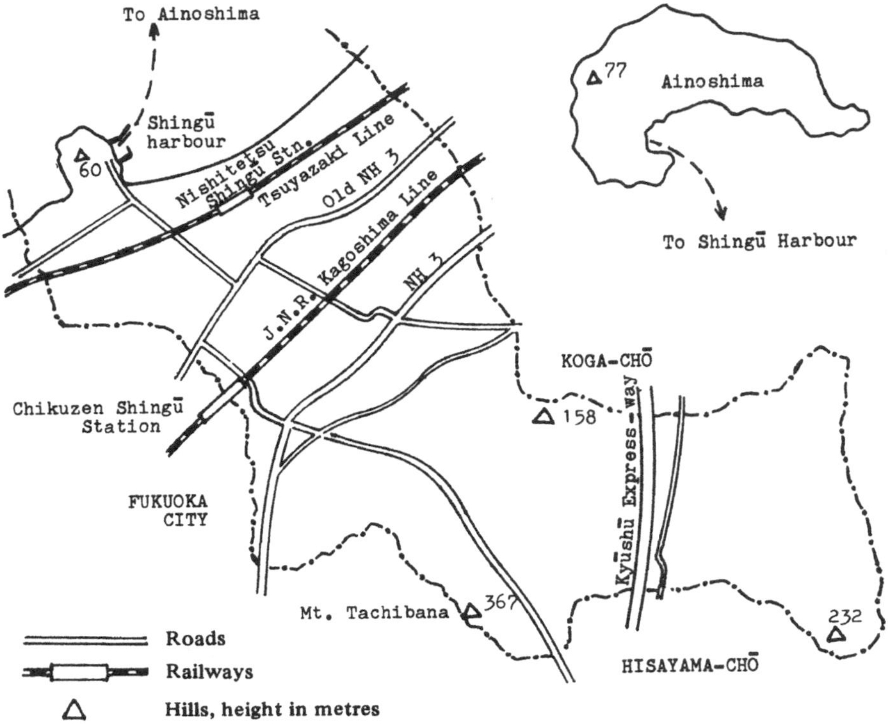

Map 2 The main communications in Shingū-machi

If we drive westwards on National Highway No.3 (NH 3), we soon pass the polluting monsters of Yawata and Tobata Steel Works, and after a while escape into small hills. On the congested highway we have plenty of time to look around; on the left hand side (to the south) the hills climb steeply. From these hills valuable coal, on which the steel works were founded, has been extracted. Between the hills and the sea to the right are plains separated by small forest-clad hills. The small glimpses we can get of rural charm are soon distorted by new suburbs being erected. About 40 km from Kitakyūshū City the new Kyūshū Expressway starts and will take you to Kumamoto. The expressway will eventually be connected with the one from Honshū now terminating in Kitakyūshū City, making it possible to drive on expressways all the way to Tōkyō. We do not enter this road, however, but continue straight for a few more kilometres. Then we enter Shingū-machi – this long and narrow township which starts in the hills to the south and ends on an island half an hour away by boat to the north (Map 1).

We have been approaching Shingū-machi from the east, but bordering Fukuoka City and being only 16 km from the commercial centre of that big city, most of the visitors come that way. Fukuoka International Airport connects the city with all major centres in Japan as well as with South Korea, China, Taiwan and Hong Kong. Since March 1975 Hakata Station has been the terminal of the *Shinkansen* express train covering the 1,069 km distance to Tōkyō in less than seven hours. The com-

Ill.2 The Nishitetsu tramway. This tram is by far the most convenient means of transportation for the people in Shingū. The hill in the background is Mt. Tachibana, where a castle was situated four centuries ago. Strawberries, an important winter-crop, are grown in the vinyl green-houses in front.

munications between Fukuoka City and Shingū-machi are good as several main thoroughfares cross the township. Starting near the hills to the south we have Kyūshū Expressway, but there is no interchange in Shingū-machi (Map 2). Moving northwards we then cross the new NH 3 before we reach the double-tracked trunk line operated by Japan National Railways (JNR) which connects Nagasaki and Kagoshima with Honshū via the cities of Fukuoka and Kitakyūshū. Only the slowest of the local trains between the two cities stop at Chikuzen Shingū Station, however. Further to the north the old NH 3 is winding its way through what we may call the centre of the township. Buses run regularly along this road. Furthest to the north – running through the pine-forest planted to protect the villages from the sea – we find the single-tracked tram connecting Tsuyazaki-chō to the east and Kaizuka in Fukuoka City to the west. From early in the morning till late in the evening this tram stops at Nishitetsu Shingū Station every thirteen minutes. For the fishermen in Shingū this tram is by far the most convenient means of transportation to and from Fukuoka City.

Shingū-machi is then well connected to the outside world, and its proximity to Fukuoka City has important consequences as we shall see later. The communications inside the township are on the other hand poor. There are no buses linking the different parts of Shingū-machi, and there are only four departures a day (five during the summer season) by boat to Ainoshima. The first boat leaves Ainoshima at 7.00 a.m. and the last leaves Shingū harbour at 5.30 p.m. making it difficult for many people to commute.

The Sub-division of Shingū-machi

Shingū-machi is today divided into nine *ōaza*, one of them is Ōaza Shingū (Map 3). The town is also divided into fourteen wards or *chiku*. In most cases the *ōaza* and the *chiku* are identical. This is the case regarding Ōaza Shingū and Shingū-ku, as already indicated. Only Shimonofu and Kaminofu are divided into several wards.

A smaller unit is the *aza* of which there are 361 in the township. In Shingū-ku there are two: Yashiki and Kita-no-guchi. The vast majority of the houses are located in Yashiki. Kita-no-guchi consists mainly of common land like the beach, the shrine and temple compounds, the community hall etc. The *aza* unit is in Shingū-ku only used for references regarding titles to land, and it does not carry any meaning in formation of groups or in local administration. The smaller neighbourhood groups (*tonarigumi*) do for instance cross the borders of the *aza*.

The neighbourhood group is of much greater importance, and Shingū-ku has nineteen of these. It is a formalised group with a chairman and occasionally a cashier. The smallest in Shingū-ku comprises eight households, the biggest twenty-four.[1]

One unit perhaps unique to Shingū-ku should be mentioned briefly here. Shingū is laid out in three parallel streets running in the east-west direction. These three streets are called Honmachi, Nakamachi and Uramachi (Map 4). Each street forms an informal and extralegal unit called *nagare* which has some importance in the policy of the ward as well as for people's sentiments. The neighbourhood groups along one street cooperate in different ways, and I will return to this later.

The administration of the township can be regarded as belonging to two levels. One is the formal administration linking the town with the prefecture and the nation. The township is the smallest legal unit in the administration. The town council is elected every fourth year, the major being elected separately. Most of the

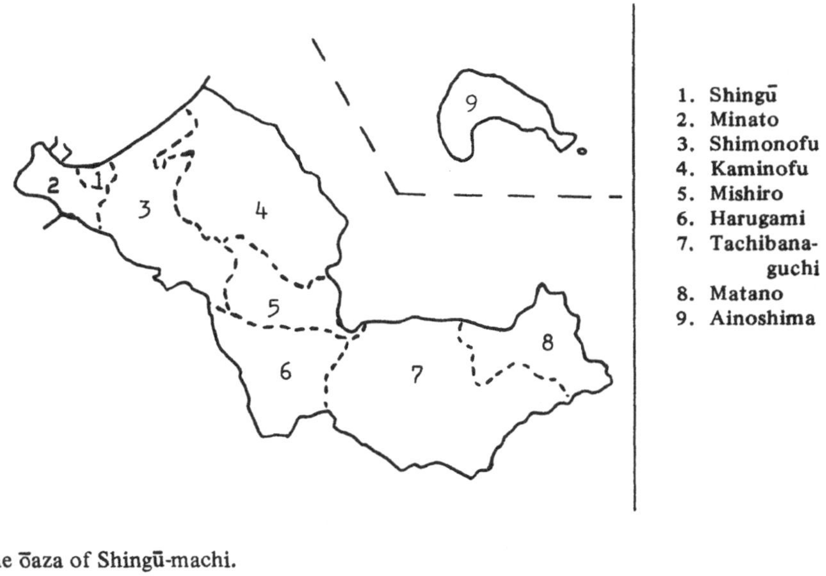

Map 3 The ōaza of Shingū-machi.

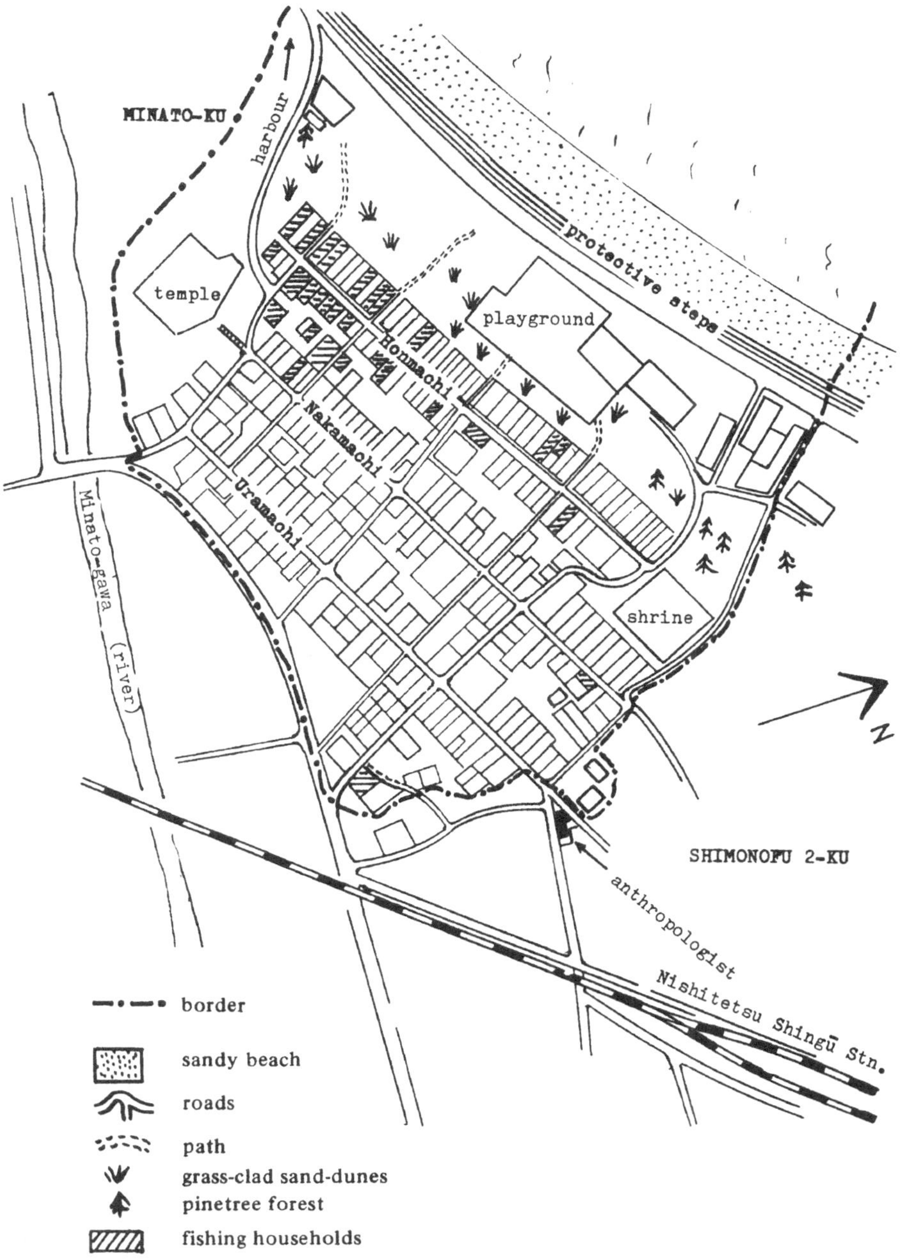

Map 4 Shingū-ku

candidates run independently, but in actual policy the majority sides with the ruling Liberal Democratic Party. The two candidates elected from Shingū represent, however, the Japanese Communist Party and the Buddhist party Kōmeitō. (This does not mean that the voters support these parties.[2]) The other level links the household (*ie*) as a member of the neighbourhood group to the town office via the ward level. In Shingū-ku three executives are elected for a two-year period, and these serve as intermediaries between the administration and the chairmen of the neighbourhood groups.

There are three elementary schools in Shingū-machi reflecting another subdivision of the township. What once was the village of Tachibana-mura has still its own school, and the children from Mishiro, Harugami, Tachibana-guchi and Matano attend this school. The children from Kaminofu, Minato, Shimonofu and Shingū attend the central school located near the old NH 3, while the children of Ainoshima have their own school. There is one central junior high school with a branch on Ainoshima. Ainoshima is still something apart, though the island was incorporated into Shingū-mura almost a century ago. In 1976 the first senior high school was opened, located in Kaminofu.

The Climate

Japan has four distinct seasons, a fact the Japanese never fail to stress. In few places – if any – are the seasons praised in the arts as in this country. The climate is remarkably cool considering the latitude. Situated just below the thirty-fourth latitude north, Fukuoka has a rather cold winter with occasional snowfalls. The winter monsoon brings cold winds from Siberia across Korea and the Japan Sea to the northwest coast. The cold winds, coupled with inadequately heated houses, make one shiver through the winter months from December to March. The Genkai Sea just outside Shingū is rough, and there are few islands protecting the coast. Consequently, there are many days when the fishermen have to stay at home, and many seek other work.

Though the plum trees are in full bloom in February, the people have to wait another month before the spring arrives. By April most are back on their fishing boats. The summer is warm and humid; the temperature may exceed 35°C. The winds are mostly from the south and the sea is usually calm, but even in this season there are days when the waves are too high for fishing. By October the temperature is again below 20°C, and more days are spent at home.

Although Japan is said to have four seasons, the picture is not complete if I stop here. In June there is a brief rainy season with heavy rains causing landslides in many parts of Japan. More important to the fishermen are the typhoons that hit the country from August to October. Developing in the South China Sea, the typhoons move north either to China or to Japan. There are two "typhoon highways" affecting Japan; one across the southwestern tip of Kyūshū turning eastwards to Shikoku and Honshū, the other across West Kyūshū to Korea. Thus, one passes east of Fukuoka, the other west. Shingū – protected by the mountains – is seldom hit directly, but strict precautions are taken every time a typhoon is approaching, and the fishermen's fear is real. As late as 15 September 1978,

Ainoshima was severely affected by a strong typhoon. A dozen fishing boats sank, and all the fish kept for breeding escaped; the damage totalled about eighty million yen (about 420,000 dollars).

Precautions have to be taken the whole year to avoid accidents. Shingū has been fortunate; the last fatal accident involved a man drowning during an epileptic attack. In April 1975, however, a boat capsized and a man drowned in Tsuyazaki, and in 1948 Ainoshima was hit by a catastrophe when several boats capsized just outside the harbour and twenty-eight fishermen lost their lives.

The Japanese find the spring and autumn very pleasant. The temperature is comfortable and the skies are all blue. The summer in Shingū is regarded as too hot, and the daily greeting goes: "Puhhh, it's hot, isn't it?". The winter, on the other hand, is considered too cold which is expressed by a shivering "Uhhh, it's cold, isn't it?".

Natural Environments

Shingū is located on a wide open bay (Map 1). The long sandy beach is framed by cliffs on either side. On the southwestern side the sixty-metre-high cliff, Isozaki-bana, protects Shingū harbour from the southwesterly winds. Four small rivers cross the beach, and four townships have developed around these rivers. The townships are Tsuyazaki, Fukuma, Koga and Shingū; all of them are old fishing villages except Koga. A narrow plain separates the beach from the hills and the mountains in the interior of Kyūshū; the highest hill in Shingū-machi is Mt. Tachibana, rising 367 metres above sea level (Ill.2). As already mentioned, the mountains protect the bay from the southerly winds, but nothing protects it from the high waves of the Genkai Sea.

The dwellings of Shingū are protected from the sea and drifting sand by a 300 metre-long concrete wall, which today is covered by sand dunes. The compounds of the houses closest to the wall are filled with sand, and Abe Kobo's famous novel, *The Woman in the Dunes,* comes to mind. A pine-forest was planted along the beach further to protect the village, but parts of it are gone after attacks by insects. It is now necessary to spray the forest by helicopter twice every summer. The whole beach and forest is regarded as a quasi national park; it is State property and it is strictly forbidden to cut trees or branches.

The river Minato-gawa passing through Shingū-machi is a muddy little creek, but it has not always been like that. Once the river carried clear water, and fish and shrimps could be caught in abundance. The 1976 April issue of *Kōhō Shingū-machi* (a monthly informative paper published by the town office and delivered free to all households in the town) could, however, report a mass death of fish at the river mouth.

Hardly a country in the world is facing the same threats to the environments as Japan. Being densely populated and with her economy aimed at reckless industrialization, the ecological destruction has been immense, and nobody has been affected like the fishermen. Many fishing villages have been forced to give up fishing altogether due to pollution and land-reclamation. Oil-spills and other wastes from industry and private households, in addition to large quantities of chemicals used in agriculture, have caused serious cases of pollution and frequent

Ill.3 The beach seen from Shingū. Huts for fishing equipment in front.

occurrences of "red tide".[3] Several bays are declared dead and the Inland Sea is in danger. But nothing can be compared to the horrors in the Minamata area in Kumamoto Prefecture, where more than one hundred persons have so far died due to mercury poisoning of fish.[4] A second case – also involving many deaths – occurred again in Niigata Prefecture shortly afterwards.

The problems of the Shingū fishermen may seem small in comparison, but to them the pollution of Minato-gawa is serious enough. Equally polluted was the river in Koga-chō, where a pulp factory emptied wastes into the river, causing the fishermen along the beach to demonstrate already twenty years ago. (The factory later went bankrupt.) Furthermore, many species of fish and shells have disappeared due to pollution and over-fishing. This might have been the fate of the sardine and sand lance, traditionally the two most important species in Shingū.

The Minamata catastrophe had effects on the fish prices which, for a time, went down. Today the fish caught near the land often receives lower prices in the markets than the same kind of fish caught far from the coast, affecting the fishermen who fish along the coast as the Shingū fishermen do. Furthermore, the cases of pollution have threatened the coastal fisheries and many, also among the fishermen themselves, doubt the future of these fisheries, and this again influences the recruitment among the young.

Financial problems have forced many fishing cooperatives to seek income from peripheral activities. Some have earned money by selling fishing grounds for land-reclamation projects. In Shingū sand has been sold for many years to construction companies erecting new buildings in Fukuoka City. Some of the fishermen claim that this practice is causing damage to the breeding grounds for fish and shells; this is a problem complex I want to take up again in Chapter VIII.

The natural environment is threatened in many ways, but some improvements have also been made. The one most important to Shingū is the improvement of the harbour facilities. The community lacks a natural harbour, though the beach was very convenient for landing the boats in the old days. As the engine was introduced around 1925, the boats became too heavy, and increased usage of the existing harbour was necessary. Extensive improvements have been made. The breakwaters and the pier were improved, a slip with motor winch was constructed, and a building with four rooms was built. The building contains a rest room, a room for storing equipment, a freezing room for ice and a refrigerated storage room for fish. The harbour was deepened in 1975, but it still puts an effective limit on the boat size. It is unlikely that much bigger boats than the present maximum of 4.5 tons can be harboured in safety.

Another shortcoming of the harbour is the direction of the waves and the tides which often hit the boats side-on when they leave and arrive. This was stated by many as the reason why the Shingū fishermen had to return soon if the weather turned bad. Among the fishermen from other communities they were regarded as cowards.

Scrap-iron and concrete blocks have in many places been dropped purposely into the sea to form *sakana no apaato* (lit. fish apartment). These *apaato* encourage the growth of plankton, which is food for small fish and shells which again become food for large fish. An extensive usage of this strategy in Shingū is ruled out as they hamper the operation of trawl-like equipment used by most fishermen in Shingū.

Ill.4 The mouth of Minato-gawa. According to Kaibara Ekken boats were harboured here in a bay several centuries ago, until the bay was filled by sand.

Some Important Species Caught in Shingū

The Japanese archipelago is located where the warm current, Kuroshio, and the cold currents, Oyashio and Riman, meet (Map 5). A branch of Kuroshio named the Tsushima current enters the Japan Sea through Tsushima Strait between Korea and Japan, and this affects Shingū profoundly. These currents provide Japan with rich fishing grounds, among the very richest in the world.

A great number of species are to be found; even polar fishes in the north and tropical ones in the south. The existence of a species does not necessarily imply that it is sought by the fishermen, however. Only when there is a demand for the product does the species become a niche for the fishermen. A niche in turn can only be made available by the use of certain technologies. The species caught, while not exactly determining the technology used, puts certain constraints on the equipment and thus the social relations of production. Thus a fish like the sardine that comes in large shoals is best caught by large drag seines or purse seines, both requiring a large number of people, while demersal species may better be caught by hook-and-line operated by a couple of fishermen.

A full account of the species found in Shingū would be tedious and to no purpose, but it is necessary to introduce the more important ones.[5] The species exploited by the fishermen in Shingū can conveniently be divided into four categories; migratory (pelagic) fish, demersal (stationary) fish, molluscs and sea plants.

The migratory species were previously by far the most important, and sardine (*iwashi*) was until recently the single most important fish in Shingū. There were several kinds; the two most important to Shingū were *ma-iwashi (Sardinops melanosticta)* and an anchovy *katakuchi-iwashi (Engraulis japonica).* Both come in shoals and were hunted mostly during the autumn and the spring. Another important species used to be the sand lance or *kanagi (Ammodytes personatus)*

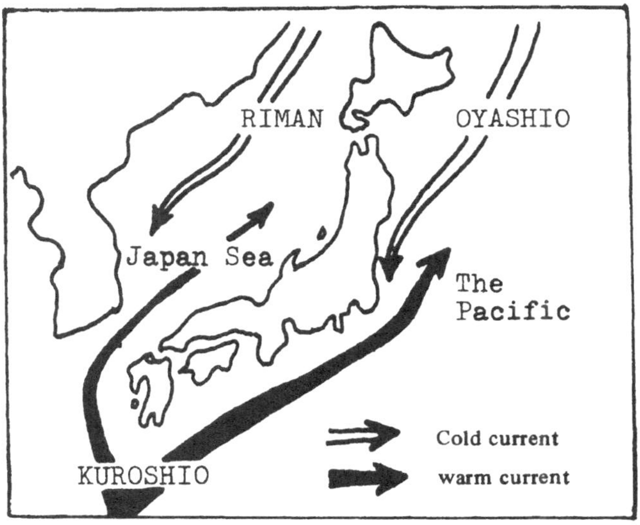

Map 5
Sea currents passing Japan.

that traditionally was second only to sardine. Other species of some importance at present or in the past are yellowtail *(Seriola quinqueradiata)* which is called *buri* or *yazu* in Shingū, *kanpachi (Seriola purpurascens), hirazu (Seriola aureovittata),* horse mackerel or *aji (Trachurus trachurus),* mackerel or *saba (Pneumatophorus japonicus),* Spanish mackerel or *sawara (Scomberomorus niphonius)* and grey mullet or *bora (Mugil cephalus).*

The migratory species have declined in importance since World War II, and today most of the sale in Shingū derives from two stationary species, the sea bream *(tai)* and a fish called *kisugo* locally or *kisu* in standard Japanese. *Kisugo (Sillago sihama)* is found on muddy grounds with small stones, and has been hunted since the 1950s, when new technologies were developed. *Kisugo* fetches high prices at the markets, as do the sea bream of which there are two kinds: *madai (Chrysophrys major)* and *chidai (Evynnis japonica). Madai*, which is the most important of the two, spawns during the spring and by summer the fry has reached a size of about 5 cm. This fry has become the niche for a short but very profitable fishery in the summer. Sea bream has traditionally been caught fully grown, but this fishery has also declined. A black porgy called *chinu (Mylio macrocephalus)* was also of some importance in the past, as were pig fish or *issaki (Parapristipoma trilineatum)* and a gizzard shad *konoshiro (Clupanodon punctatus).*

A wide variety of octopus *(tako)* and squid *(ika)* are caught in Shingū. They are present more or less the whole year, but are caught mostly during the winter and spring for lack of more profitable species. A few families are catching crabs, a few varieties of sea cucumber *(Holothroidea)* and shells, but these are only marginal activities with little overall importance today. One species of snail should nonetheless be mentioned, namely the Japanese ivory shell or *bai (Babylonia japonica),* which is a snail living on sandy shallow grounds. It has been left alone until a few years ago, when a profitable market developed. It has today a very high value, and several boats are engaged in catching it. Though it is available most of the year, it is mainly caught during the winter when it is the most important species for many.

The Japanese are well-known for eating many kinds of sea-weed, and some fishing villages specialize in cultivation of sea-weed. There are four important kinds; *wakame (Undaria pinnatifida), nori (Porphyra), konbu (Laminaria)* and *tengusa (Gelidium).* Of these only the first had any importance in Shingū.

Wakame can be collected near land during the spring, but this plant is also suitable for cultivation. The seeds are "planted" in November and the harvest begins at the end of February and continues through May. Three boats had equipment for *wakame* cultivation, but they ceased cultivation in 1977.

The fishing cooperative tried some years ago to introduce *nori,* but the experiment failed. It was found that the sea was too rough. It is, on the other hand, extensively grown in Hakata Bay and in Tsuyazaki. Neither *konbu* nor *tengusa* are found in Shingū, but *egonori (Campyraephora hypnaeoides)* resembles the latter. *Egonori,* which is locally known as *okyūtogusa,* is found in the Japan Sea and the Inland Sea and can be mixed with *tengusa* to make gelidium jelly.

The great variety of species in the waters of Shingū has caused a considerable variation in technologies, but before we turn to this subject it will be necessary to outline the history of Shingū and the social organization of the community today. These will be the themes in the next two chapters.

CHAPTER III

THE HISTORY OF SHINGŪ

The history of Shingū can conveniently be divided into three epochs; the time before, during and after the Tokugawa period. Hardly anything is known specifically about Shingū before the emergence of the Tokugawa shogunate; in fact Shingū did not come into existence at its present site before well into that period. But the community was not created from nothing, the seeds were sewn a long time before. To understand the founding and the subsequent development of Shingū, a few words will be said about the earliest times.

The History before the Tokugawa Period

Shingū has never been a remote isolated village. On the contrary, the site has often been in the centre of important events. Northern Kyūshū served as a gate for innovations brought to Japan from the mainland, often directly from China but also through Korea. The single most important innovation was the technique of wet-rice cultivation brought by immigrants about 300 B.C. This marks the beginning of the Yayoi period (200 B.C. – A.D. 250). The Jōmon period (before 200 B.C.) knew only more primitive forms of agriculture, but the people of that time relied mostly on food-collecting, fishing and hunting. Huge dunes of shells called *kaizuka* – of which Shingū-machi has its share – indicate settlement patterns as well as eating habits. During the Yayoi period new people are believed to have arrived at Kyūshū, and the new inventions spread to Southern Kyūshū and Honshū.

Kinki (including Nara, Ōsaka and Kyōto) emerged in the beginning of our era as the political centre of the nation, but due to its strategical location, Northern Kyūshū ranked second. A regional capital, Dazaifu, was founded both to administer the island and to handle foreign affairs. Dazaifu became the second town in Japan, second only to the capital. The highway connecting the capital and Dazaifu became the main road of the empire. This highway passed near Shingū, and the nearest station where horses, provisions and accommodation were available was Aoyagi, near Shingū in present Koga-chō. The administration was built on the Chinese model, communications with China were frequent, and parts of Korea came under Japanese rule. All increased the importance of Northern Kyūshū.[1]

Towards the end of the Heian period (794–1185) the centralized administration decayed, and powerful lords emerged all over Japan. Local fiefs developed, the most powerful being those of Taira and Minamoto. These clans were engaged in a long and bloody struggle which ended with the sea battle in 1185 on the strait between Honshū and Kyūshū. Remnants of the defeated Taira fled and spread over large areas, also on Kyūshū.

The period from the establishment of the Kamakura shogunate (1185) to the emergence of the Tokugawa shogunate (1603) was an unstable one, and several powerful houses were fighting each other in Northern Kyūshū. The control over Chikuzen province where Shingū was located changed hands several times during these four hundred years.

Minamoto Yoritomo tried to pacify Kyūshū after his victory over the Taira, and he established several new offices and appointed military commissioners or *shugo* to all the provinces. Two of the offices he created in Kyūshū were the *Chinzei-bugyō* or Governor General and the *dazai-shōni,* a term Sansom translates as "junior assistant" (1974a:443n). Mutō Sukeyori, a descendant of Fujiwara Hidesato, was appointed to the latter office, and the family took the name Shōni. Sukeyori was also appointed *shugo* for Chikuzen, Buzen and Hizen provinces and Iki and Tsushima islands, and emerged thus as a powerful lord in Kyūshū. During the Ashikaga period (1338–1573) Shōni Yorizumi was appointed *Kyūshū tandai,* an office that replaced the *Chinzei-bugyō.* But Shōni had strong enemies, particularly Kikuchi in Higo and Ōuchi in Nagato across the Kannon Strait. Chikuzen was taken by Ōuchi and the last Shōni committed suicide in 1556 after two defeats by his vassal, Ryūzōji.

One of Minamoto Yoritomo's most faithful followers was Ōtomo Yoshinao, descendant of Fujiwara Hidesato and the founder of the powerful Ōtomo family. For his support he was appointed *shugo* of Bungo in 1193 and Governor General in 1223, a position that became hereditary (Papinot 1972:500). His descendants increased the domain, and at the time of the fifth generation they ruled Bungo, Buzen and Chikugo provinces. A branch founded by Ōtomo Sadatoshi established its castle on Mt. Tachibana in present Shingū-machi where Sadatoshi built the Tachibana Castle in 1330 and took the family name of Tachibana. The castle came to play an important role for 250 years, and several wars were fought at the foot of the mountain.[2]

Ōtomo had long been involved in struggles against the Shimazu family in Satsuma and Ōuchi (from 1554 replaced by Mōri) in Nagato. In 1584 Ōtomo Sōrin (alias Yoshihige) was forced to appeal to Hideyoshi for help against Shimazu. Hideyoshi's great Kyūshū campaign took place in 1587, and Shimazu was defeated. Nevertheless, the domain of Ōtomo was reduced, and the family retained only their original province, Bungo. Tachibana Muneshige, who defeated Shimazu in 1586, was given Yanagawa fief in Chikugo, and Kobayakawa Takakage, third son of Mōri Motonari, was given Chikuzen province including Tachibana Castle in 1588. Another of Hideyoshi's able commanders, Kuroda Yoshitaka, was given parts of Buzen province where he put up his castle in Nakatsu.

After the death of Hideyoshi, Tokugawa Ieyasu moved to take the power, and among his supporters were both Kuroda and Kobayakawa. After the victorious battle at Sekigahara in 1600, Kuroda Nagamasa received Chikuzen and Kobayakawa Hideaki was given Bizen and Mimasaka provinces.

In the latter half of the sixteenth century the Christian faith arrived in Japan through Kyūshū where it gained a stronghold. Several of the lords became converts, more for political than religious reasons. With the Portuguese and the Dutch who brought this new faith, came also new and modern weapons, and Ōtomo was the first to use cannon in battle (Hall 1971:138). One of the converts was Ōtomo Sōrin who was baptized in 1578, and four years later he sent together with other Christian lords Japan's first envoy to the West (op.cit:140). Sōrin tried "actively" to spread Christianity in his domain, including Shingū. It is unknown whether he had any success. More important was the revival of Buddhism in this period with establishments of many new temples. One of these was Sainen-ji of the Seizan Jōdō sect in Minato-mura. It was established in 1598, and subsequently all the Shingū fishermen became members of its congregation.

Shingū during the Tokugawa Period (1603-1868)

Towards the end of the sixteenth century three strong leaders emerged from the chaotic mess of civil wars, and Japan became once more united through the efforts of Oda Nobunaga, Toyotomi Hideyoshi and Tokugawa Ieyasu. The shogunate founded by Ieyasu lasted for 250 years, a period that perhaps is most famous for the isolation it brought to Japan. The country was almost sealed from foreign contacts, only Nagasaki was open to Dutch and Chinese traders. (Actually some Korean missions also arrived in Japan during this period, and they often landed on Ainoshima. Both Nagasaki and Ainoshima had, as I will show later, some importance to the fishermen in Shingū.)

To consolidate their position the Tokugawas reallocated the provinces, and Kuroda received Chikuzen. This province was divided into Fukuoka-han (where Shingū was located) with a tax-revenue (*kokudaka*) of 520,000 *koku* and Akizuki-han (50,000 *koku*) where a branch family of Kuroda became lords.[3] The warriors who previously partly had been living in the countryside, were removed to the castle towns and thus alienated from the land. Many of these former warriors, however, chose to remain in the countryside where they became leading farmers and merchants, and two leading houses in Shingū claim such background.

Furthermore, the Tokugawa enforced a rigid social system of stratification which by many have been described as a caste system. The society was conceived as consisting of four classes; the warriors, farmers, artisans and merchants in that order in the hierarchy. Below these we find the outcastes: *eta, hinin, ebune* and others. Everybody was classified according to class, and in the process many warriors were reduced to farmers and merchants.

In 1640 forced registration at a Buddhist temple was introduced in order to ensure that nobody believed in the prohibited Christian faith. The population of Shingū became mostly congregational members (*danka*) of Sainen-ji.

It was in this atmosphere that Shingū was born, or more correctly was moved. From old times there was a bay in the lower part of the river Minato-gawa floating through the township, and this bay formed a natural harbour. The fishing village was located to the west or southwest of this bay. The name Minato (Lit.: harbour) of the area even today also indicates that there was a harbour on the site. As the centuries passed, the bay was due to the winds "gradually reclaimed without people's notice and became rice fields". The famous Japanese Confucianist Kaibara Ekken (1630–1714) continues in his description of Chikuzen province:

> (The West bank) was the fishermen's living quarter and it was far from the beach and very inconvenient for fishing. That is why during the Tenna era (1681–84) the beach public officers . . . were kind towards the fishermen, levelled the ground and prepared it for houses, built the earth-wall outside, planted pine trees and moved the houses of the people to the beach (Ekkenkai 1973:438-39).

It is, however, hard to believe that Shingū was moved solely to please the fishermen, the more so since the village was moved only a few hundred metres, and it took a considerable amount of money and manpower to develop the new site. The relocation was desired apparently for other reasons on which we can only make some guesswork.

The first factor in the desired relocation was the administrative point of view. During the Tokugawa period a number of attempts were made to improve the efficiency of the bureaucracy in order to control the populace. One of the measures was to separate the different segments of the population, both geographically and administratively. Chikuzen was administratively divided in three, and each division was headed by a *bugyō* or magistrate. There was one *machibugyō,* one *gunbugyō* and one *urabugyō* administrating the towns (Fukuoka and Hakata), the farming villages and the fishing villages respectively. Shingū-ura was geographically located in Kasuya County, but not administratively. All the fishing villages in Chikuzen were organized under the *urabugyō,* and the villages were divided into districts *(fure).* The number of districts varied, but towards the end of the eighteenth century was reduced and fixed at three. Shingū-ura was included in the central district comprising the fishing villages in the counties of Kasuya and Munakata. The district headman *(fureguchi* or *ōjōya)* was the link between the military administration (represented by the *urabugyō)* in the castle town and the village headmen *(shōya)* of his district. By separating the fishermen in Shingū from the farmers in Minato and by giving Shingū status as a fishing village, the two segments of the population were firmly separated. Only the temple was shared.

The farmers and the fishermen had different duties *vis-à-vis* the ruling class, and this might be a second reason making the relocation desirable. The farmers had to pay taxes in rice, and all the farm land was classified in a survey in order to estimate the tax revenues of the lords. Similarly, the fishing villages were given fishing rights over certain territories, but were in return obliged to pay taxes both in kind, cash and corveé labour in the form of *kako* (also called *funako;* crew member or worker on a boat) according to the size and economic standing of the villages. The duty of *kako* was a heavy burden on the fishing villages, and a number of documents are applications for a reduction of the number of *kako* (cf. note 9, this chapter). It seems that the *kako* of Shingū had five duties: (a) The most important duty was to protect Nagasaki harbour from foreigners, and to transport officials between Fukuoka and Nagasaki. (b) To transport officials to and from Ainoshima, a duty which was particularly burdensome just before and during visits of Korean missions to the island.[4] (c) To participate during the transport of the lord and his retainers when they left for or returned from Edo.[5] (d) To load and unload ships when these anchored in the harbour.[6] (e) To take care of and rescue drifting ships outside the beach.[7] It was thus necessary for the authorities to secure the services of the *kako,* and this was best done if the farmers and fishermen were separated. It would be simpler to calculate the number of *kako* available and to check that the determined number of *kako* was sent. The main duty of the district headman was to provide the authorities with all necessary data on the population and the wealth of the fishing villages and to ensure that the ordered number of *kako* was made available.

Finally, the relocation of Shingū must be seen from the merchants' point of view. A large proportion of the households in the village at that time were merchant households. A document from 1692 states that only fifty-eight households were engaged directly in fishing while eighty-four households were engaged in different

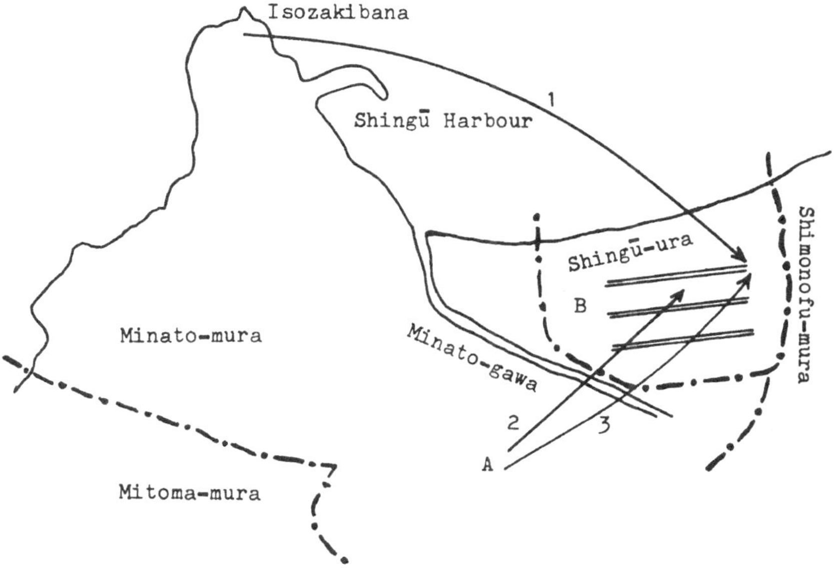

Explanation:

A: Old location in Manato-mura, aza Furuyashiki.
B: New location in Shingū-ura.
1: Relocation of Isozaki shrine.
2: Relocation of the fishermen.
3: Relocation of Haenominato shrine.

Map 6 The old and new locations of Shingū.

businesses. Twenty-three houses were empty (Kaneuchi Doc.129; in Kitamura 1976). One reason for this high number of merchants is probably the important traffic to Ainoshima (Kitamura 1976:6). Moreover, the relocation coincided with a rapid growth of commerce in Japan that "reached its height in the famous Genroku era (1688-1703) and from then on maintained itself on a high level throughout the rest of the Tokugawa period" (Sheldon 1958:4). Shingū might have been a rather flourishing commercial centre at the time, a fact that harmonized badly with the prohibitions against merchants living in rural communities. Rural commerce was suppressed when possible, and it might have been easier for the merchants to work in a fishing village where the restrictions were relaxed (Yazaki 1968:164). The relocation might thus have been desired from some of the merchant houses, and one of these, named Mantokuya, seems to have grown rapidly after the relocation.

The exact year of Shingū's relocation is a little uncertain, but 1685 has been regarded as the most likely year. In that year two shrines (Isozaki-jinja and Haenominato-jinja) were moved from Minato to the present location, and the first protective wall was erected about the same time (Map 6). A small harbour with breakwaters was constructed a little later mainly to serve as the harbour for the transport to and from Ainoshima, perhaps also to harbour Mantokuya's allegedly commercial ships, and finally as an emergency harbour for the fishermen.[8]

There is a possibility that the movement to the new location was done step by step, and that some of the fishermen put up huts along the beach before 1685. These huts were originally erected for storing equipment, but have then gradually been converted to living quarters. The first settlement is said to have been in the northwest corner of Shingū, a corner that became known as Kitagata (lit. North side), as it was located north of the old site. By studying the temple books (where all deaths should have been recorded, but which unfortunately are incomplete at this time), we can get some idea about the movement into Shingū. The first recorded death in Nakamachi is in 1685, i.e. the same year the village was moved. No death has been recorded from Honmachi before the next century, however. It is therefore reasonable to assume that the term "Hama" (lit. beach) has been used for this street, and the first death in Hama is registered already in 1661. If this is so, the first fishermen moved before the village was officially relocated.

It is on the other hand certain that by 1701 Shingū-ura was firmly established on the new site, and a map indicating the borders to Minato-mura and Shimonofu-mura is preserved from that year (Ill.5). The village was centrally planned and laid out in three parallel streets crossed by narrow lanes. The land was divided into plots so that each house in Honmachi had a facade 4.5 metres wide. This pattern is preserved to this day. The relocation of Shingū was completed in 1747 when Sainen-ji was moved to the west side of the village.

Three parties seem to have been involved in the relocation – the authorities, the merchant houses and the fishermen – and all seem to have benefited from the movement. This made the relocation smoother, of course, but it was not carried out without hardship to the fishermen. To help them house-plots were given free, and they were exempted from paying certain taxes during the first years. Nonetheless, a document dated 1692 says that the people were almost starving and had applied for reduced tax burden as *kako.*[9] Pestilence and starvation were not far away; in 1687 twenty-six children died in Hama, and in 1732-33 a terrible famine hit the whole area and between a third and a half of Shingū's population perished.[10] To help people through such disasters a fund called *tame-gin* was established, and this was partly financed by the authorities and partly by an extra tax in the villages. This fund was frequently used; in 1825 three fire victims in Shingū borrowed 855 *momme* silver which was to be paid back in five years.[11] In 1828 two typhoons hit the village and thirty-seven households borrowed 5 *kan* 550 *momme* silver for repairs, and eighteen fishermen borrowed 5 *kan* 200 *momme* silver to build new boats. And again the next year eleven fire victims borrowed 2 *kan* 887 *momme* silver (Akaishi Doc. in Komonjo o yomukai 1976). In spite of such funds most people lived in poverty, though there were differences also among the fishermen. Some fishermen did manage to accumulate some capital and owned big nets, and these might have tried to avoid the burdensome duty of serving as *kako,* so that it became necessary to stress that this should not happen, or to use the words of the Decree of 1772: "We should try to return to the old custom, i.e. that the sons of the well-to-do went as *kako.*" (Fukuoka-ken 1963:652). It can be assumed that it

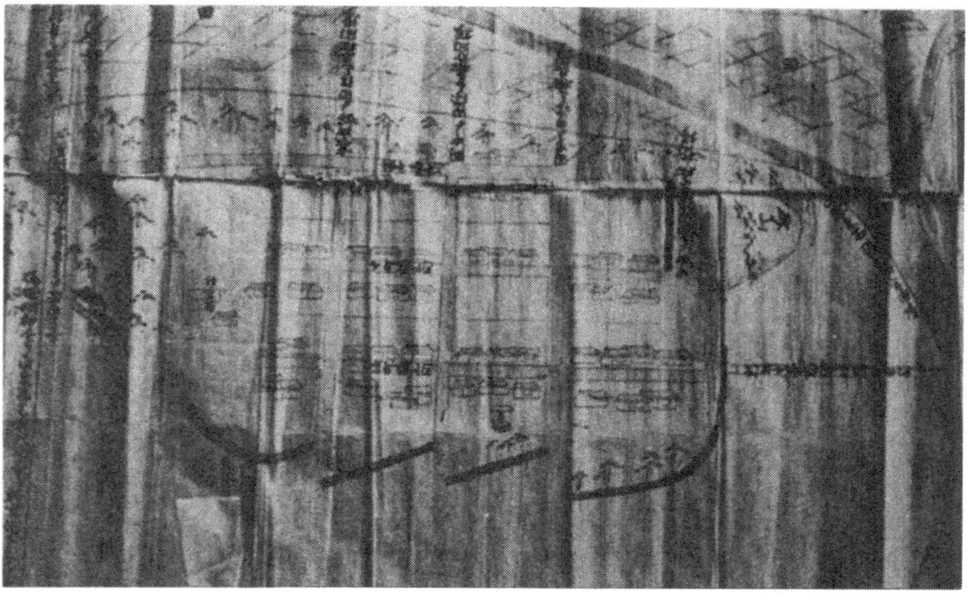

Ill.5 A map of Shingū-ura from 1701. The map was made to show the borders of Shingū-ura. The three streets are clearly seen as is the wall to protect the village from the sand.

was the poorer fishermen who actually served as *kako*, and the poor families lost workers they desperately needed. The difference between the "rich" and the poor fishermen might have increased.

During the Tokugawa period the villages became relatively autonomous, and the administration of the village was the responsibility of the village headman and the elders or *kumigashira*. The collection of taxes, execution of law and order, and the registration of the population were among the important tasks. The village headman should also walk around in his village every day to check the needs of the people, and he had to manage the budget of the village. The appointment of the headman could be done in several ways. In some villages the position was inheritable while in others this position rotated among a segment of the population or he was elected. In Shingū-ura the position probably rotated; at least it was not inherited.[12] In either case, the headman was recruited from the richest and most powerful families in the village. In Shingū these families were found among the merchants. To assist him he had some *kumigashira*, and the number of these in Shingū seems to have varied between two and five. Though the *kumigashira* should represent the population, it can also be assumed that these came from the upper classes as they had to read, write and calculate.

The district headman had to supervise the village headmen in his district, control their accounting, carry information between the castle town and the villages etc. He had also judicial power as had the village headmen. Legal cases were settled by arbitration in the villages, and the headmen tried to avoid bringing cases to the *bugyō* (Wigmore 1969:43-45). Settlements by conciliation also put the headmen in a crucial position as conciliation "simply serves the dominant power in the community" (Henderson 1965:8). The judicial power thus embedded in the positions of district and village headmen indicates an important aspect of the

Ill.6 The western half of Uramachi where merchants used to live.

fishermen's dependence as disputes over fishing grounds were frequent. Moreover, in Chikuzen the district headman held decisive power of the *tame-gin* fund, which also gave the district headman great influence over the fishermen who were dependent on this fund.

The Tokugawa period has been known for its stability and rigid control. The people were restricted both in their physical as well as social mobility. The official hierarchy with the merchants as the lowest ranking of the four classes did on the other hand not have any relevance in actual life. (See for instance Sheldon (1958: 26-31) for the discrepancy between the ideal and the real patterns.) The merchants steadily accumulated more wealth and controlled the fishing population who were – in theory – their superiors. The fishermen were strictly speaking referred to as farmers (*hyakushō*) except for small groups like the *ebune* who lived on their fishing boats. The street where most of the fishermen lived was called Honmachi, which means the main street, while the merchants lived in Uramachi (the back street) and Nakamachi (the middle street). At the same time the landlords started to develop rural industry while the warriors became poorer and poorer. Though the Tokugawa were remarkably successful in keeping the country under control for 250 years, they could not freeze the development, and the regime was ready to fall when the American Commodore Perry knocked at the closed door of Japan in 1853.

Shingū after the Tokugawa Period (1868-)

Tokugawa's concessions *vis-à-vis* Perry gave some young warriors the opportunity to challenge the regime. The young warriors of Satsuma and Chōshū had for a long time been in opposition to the central government, and these fiefs launched attacks on the foreigners in Kagoshima and Shimonoseki. Later these warriors became instrumental in the *coup d'ētāt* which became known as the Meiji Restoration in 1868. During the next century Japan changed from being an isolated feudal society to a highly industrialized capitalist power.

Shingū became integrated into the nation on new levels. The distinction between farming and fishing villages was lifted, and Shingū was integrated into Kasuya County. A county assembly was introduced and elections held from 1896. Each village could elect one representative, six landholders owning land worth more than 10,000 yen were to be elected in addition. Two of these came from Shingū-mura. The system was changed three years later when the privilege of the landowners was dropped. On the other hand, Shingū-mura was entitled to two representatives, but lost the second in 1918. Shingū-mura certainly declined in importance. Five years later the county assembly ceased to exist, and the county lost its importance in local politics.

Shingū-mura also became a village in the modern Fukuoka Prefecture, and the inhabitants participate in the elections both for the assembly of the prefecture, for the governor and for representatives to the national Diet in Tōkyō. The village is then well integrated into the national political system, not only into the feudal fief as was the case previously.

When Shingū-mura was created in 1889, the influence of the inhabitants of Shingū-ku declined. The big landholders in the new unit emerged as the powerful leaders; the position as mayor was monopolized by the Yasutake family in Kaminofu from the creation to 1916. Yasutake and the big landowner in Minato dominated as representatives to the county assembly. In 1896 Sakai Manshichi in Minato owned land worth 26,480 yen and Yasutake Usaburō land worth 11,772 yen; both were elected to the county assembly.

The merchants in Shingū-ku lost their direct access to the military administration via the village and district headmen, who were merchants like themselves. Moreover, the judicial authority was now given to the police and the courts, and the *tame-gin* fund disappeared. The merchants lost some of their assets in their transactions with the fishermen who on their side obtained more influence in local affairs. The new position as headman of the ward *(kuchō)* who took over some of the minor tasks of the former village headman, began to rotate among the streets. As Honmachi was solely inhabited by fishermen, some of them came to hold the position of headman.

New systems of transportation developed during the Meiji era (1868-1912). The Hakata-Akama railway passing through Shingū-mura was opened in 1890, but there was no station in the village before 1920 when Chikuzen Shingū Station was opened. The Sino-Japanese war caused a lot of activities around Hakata Bay, and many new railways were subsequently built. Coal mines in the area also caused a boom to the railways, and Fukuoka Prefecture is today criss-crossed by unprofitable lines. The new lines increased the opportunities for pilgrimage, and some lines were built specifically for this reason. One of them was the *Nishitetsu Miyajidake Line* opened in 1925 and later extended to Tsuyazaki. This line is probably the most convenient link to Fukuoka City for the inhabitants of Shingū. The existing roads were improved and new ones constructed. NH 3 was built through Shingū-mura, and due to the Sino-Japanese War a new road was built linking NH 3 and

Shingū harbour. Telephone and telegraph were introduced at the end of the nineteenth century, and the first post office opened in 1902.

The new infra-structure that emerged after the Tokugawa period did not favour Shingū. As more of the transport went overland, the harbour declined in importance. The railway station on the other hand was opened only thirty years after the railway started operations, and when it was built it was located in Shimonofu, which has developed as the new centre of the township. Shingū became a peripheral community, and the service available locally decreased. The school that first opened at Isozaki shrine in 1872, was moved to Kaminofu; Araki moved his hospital to Fukuoka City and many of the shops have closed. Shingū saw a great mobility of its merchants during the Meiji era, and there were few opportunities for work locally.

Outside the community, however, new opportunities appeared as industrialization progressed. Some left the fishing boats before World War II to take jobs in coal mines and elsewhere. The proximity to the cities of Fukuoka and Kitakyūshū, coupled with the development of the new transportation systems, facilitated work opportunities without physical migration to an urban centre. Industrialization and subsequent decline in the fishing population did not, therefore, cause an exodus as was the case in many other areas, and what is now happening to Ainoshima. This decline in the fishing population continued during the war as most young men were drafted, but the years following the war saw a temporary increase. However, since about 1950 the decline has been steady (see next section).

The social organization during the Tokugawa period was aimed at rigid social control. The population was encouraged to extreme frugality, and they were organized in small neighbourhood groups of from five to twenty households (*goningumi,* lit. five men's group). The whole group was held responsible for the behaviour of the individual, and they were encouraged to report on each other. The emergence of the Meiji era did not change all this. The old values continued on the local scene for many years. In fact these values received renewed attention in connection with the Local Improvement Movement (*Chihō kairyō undō*) introduced in the first decade of this century, and "which may be regarded as the first organized and sustained effort of the Japanese government to deal with social problems created by industrialism and imperialism" (Pyle 1973:52). This movement had three basic aims: (1) averting the social disruption seen in Western industrial societies; (2) absorbing new groups into political processes in a way that would not jeopardize the existing order; and (3) promoting continued rapid economic development (op.cit.:53).

In Shingū this movement affected the inhabitants in two ways. Firstly, local rules of "good behaviour" were stressed, and the community was given opportunities for sanctions, both economic and by ostracism, against deviations. Secondly, new organizations were established in order (1) to integrate the individuals inside the community and (2) to integrate the community in the national network.

The Young Men's Association (*Seinenkai,* also known as *Seinendan* and *Seishinkai*) was established in the Meiji era, and in 1911 Shingū-mura *seinenkai* was established with six branches (based on the wards) and 282 members. The main activities were road repairs, meetings, sport etc. (Kasuya-gun 1924). At a meeting in Shingū during the sixth month of 1909 it was stated that: "Those between fifteen and twenty-five years of age shall be members of the Young Men's Association to keep up the spirit of good behaviour, hard work and saving" (Ketsugi-roku Ōaza Shingū, § 25). A similar group was organized for the young unmarried girls (*Shōjokai*), and this was in Shingū-mura organized in 1918 with 356 members in six branches. The main activities were sewing, arranging lectures and entertaining old people. In Tachibana-mura one of the aims was to help families with members in the military (Kasuya-gun 1924).

There were youth associations during the Tokugawa period called *wakamono-gumi*, and their aims were similar to the new associations'. But *seinenkai* and *shōjakai* became integrated beyond the old village borders. Firstly, they became branches of a new central association for the new village; secondly, they were integrated into the counties. The *seinenkai* was organized on the county level in 1913 and the *shōjokai* nine years later. Other associations like the Housewives' Club *(Fujinkai)* and the Firemen's Association (*Shōbōdan*) were also integrated beyond the ward level.

To keep the inhabitants in line with "good behaviour" each neighbourhood group had a *kogashira* whose responsibility it was to ensure that the persons in his neighbourhood group behaved according to the norms. In addition there existed two supervisors *(torishimari-yaku)* in each street. Six *kyōgi-in* took care that all the formal rules decided by community meetings were kept, and economic sanctions were used to improve the performance of the individual.[13] There were also rules regarding festivals and gifts in order to curtail lavish spending.

As more people took employment outside Shingū-ku, and as many of the merchants who formerly were the backbone of the local society left the community, the social contacts inside the ward declined in importance. The men in particular became well integrated into their places of work. Many local associations have received less attention from the population, specially since the public support officially was withdrawn after the defeat in 1945. Several of these associations have difficulties in recruiting members, a membership that some years ago was regarded as compulsory. A few years ago the youth clubs became inactive, though they still exist in other wards. The fire brigade finds difficulty in mobilizing fifteen members, which is the minimum required in order to obtain economic support from the township. The fishermen, for whom the locality still carries some meaning, have been isolated. As an answer to this they have formed their own organizations, thus alienating themselves further. The fishing cooperative has become a community within the community, a problem I will return to in the next chapter.

The last hundred years have brought many changes to the fishermen; the single most important one is probably the establishment of a fishing cooperative. Other changes that have improved the conditions are the introduction of the engine, of nylon nets, the construction of a new and better harbour, and so forth. On the negative side must be mentioned the decline in some species of fish formerly important, cases of pollution, and the general decline of the fishing population, and to this last problem I will now turn.

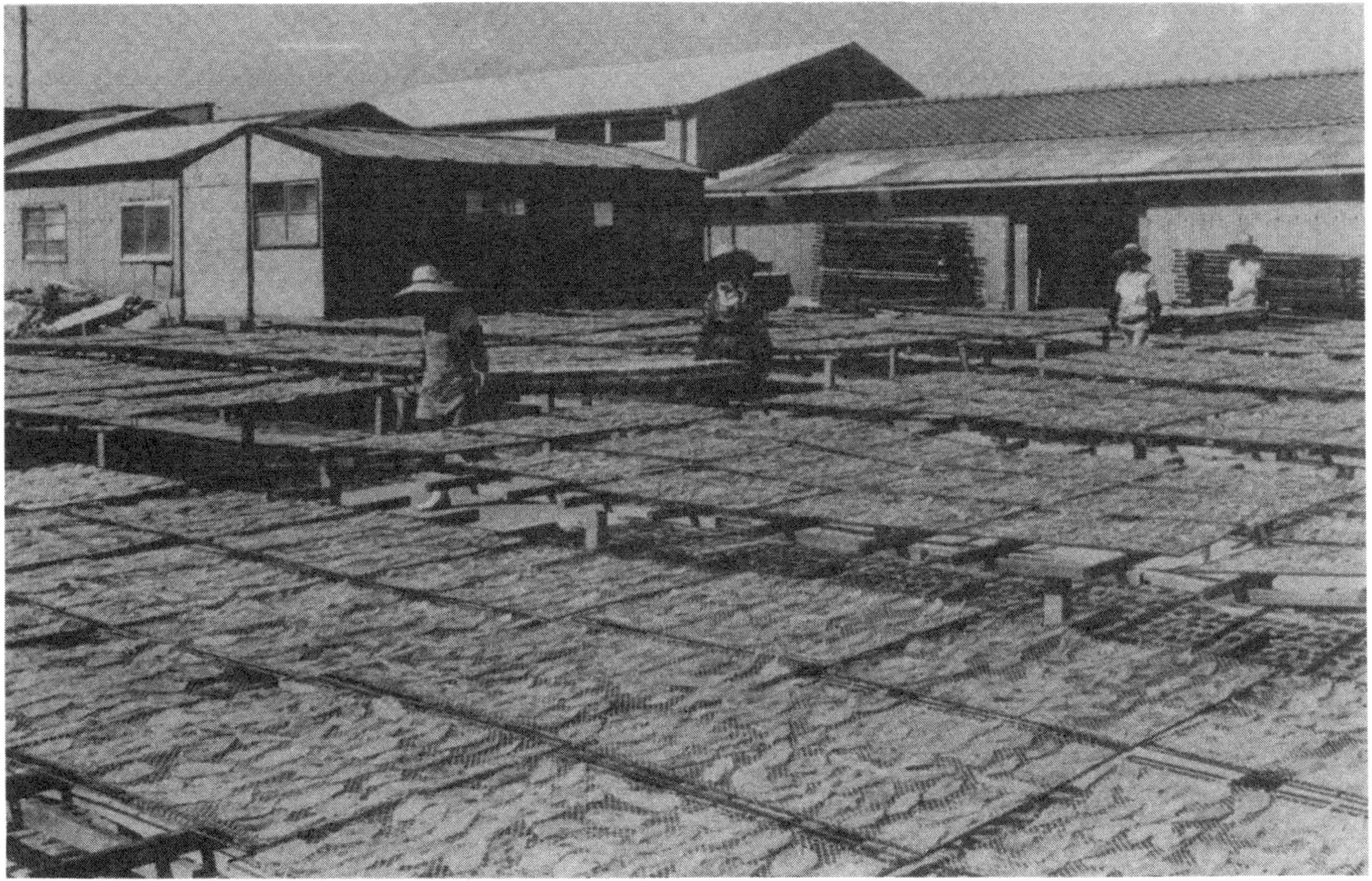

Ill.7 Kirishima Sea Product Industries. Fish is brought out in the morning to dry in the sun.

Occupational Changes

There exists a document from 1692 stating that of the 165 houses in Shingū-ura not more than fifty-eight were engaged in fishing. Twenty-three houses were empty, and the rest were engaged as *sakaya (sake*-brewing house), *kōjiya* (shop for malt, yeast and paste used for soup), *konya* (house for dyeing clothes), *rōnin* (masterless warriors), pedlars, doctors, etc. (Kaneuchi Doc. 129 in Kitamura 1976:16-18). This may seem quite remarkable since the shogunate tried to suppress the development of shops outside the towns. Only a very limited number of merchandise was allowed in the rural areas.[14] Unlike the farmers who were mostly self-sufficient (in Kyūshū cash crops comprised only 10.2% of the total agricultural product in 1877, see Smith 1959:72), the fishermen were dependent on a market-oriented economy. The fishermen in Shingū did not have any farming land; what they formerly had in Minato had been lost (allegedly covered by sand), and what some of them had in other places were only small plots for kitchen-gardens. The fish was a cash-crop, and the fishermen had to buy their staple food – rice. The fishing villages therefore often operated as small market towns. Yazaki states (1968:164):

> Controls were relaxed somewhat in the fishing villages, as it was recognized that special needs existed. An investigation made in 1782 in Okayama prompted approval of sales of lumber, fishing tackle, vinegar, soy sauce, oil and other necessities. Special conditions also yielded *sake* allowances for fishermen in Sendai in 1757 for whom, an old saying claims, "*sake* is as essential as food itself".

Though more shops were allowed in the fishing villages, it should be clear from the number of houses engaged in business in Shingū, that the market went far beyond the village itself. The importance of Ainoshima in this respect has already been mentioned, and Shingū probably served the farming villages in the vicinity as well. Many of the inhabitants recall with pride the "glory" of the past. "Shingū was the biggest town in the area, and people came from far away to shop here" was an expression I often heard, and there seems to be a bit of truth in it.

It is very difficult to obtain detailed information on the shops during the Tokugawa period, but I have made an attempt to list the more important houses in the village (Appendix II). Most of the houses were located in Nakamachi and Uramachi. Some of these merchants were of great importance to the fishermen, and in order to understand the dominant organizational form within the fisheries at that time, I will introduce in the next section the four houses that seem to have played the most important roles.

Many of the merchants disappeared from Shingū during the Meiji era due to – as we have seen – changes in the infrastructure. Others have changed occupations, and only a few shops are left in Shingū. Besides the fishing which engages about forty persons, the biggest working places are to be found in some of the fish-processing industries. These were formerly based on catches of sardine and sand lance caught by Shingū fishermen, but today all their fish is obtained from outside the township; capelan even from Norway. These factories employ mostly women who could in 1975 earn 2,000-2,500 yen per day. (There were approximately 300 yen to one US $ in 1975.) There are further two Japanese style hotels in the ward, plus one in Minato-ku on the way to the harbour.[15] There is a small factory

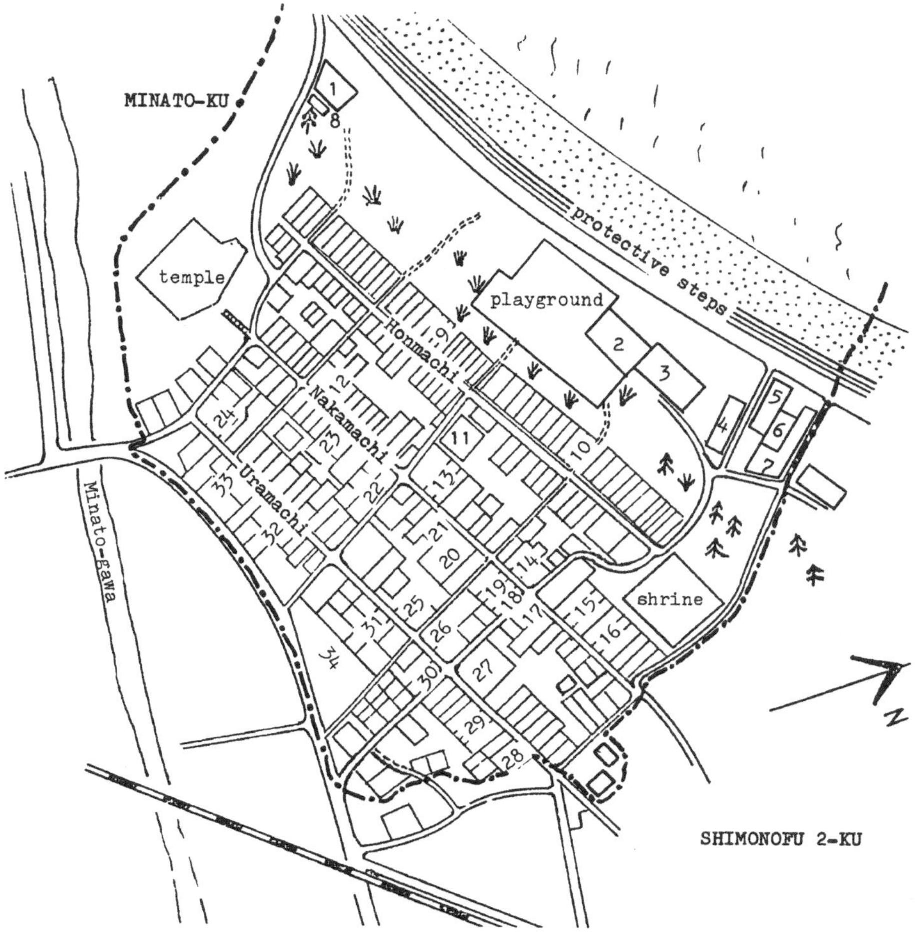

1-7	Fish processing industries	21.	Barbershop
8.	Plumber	22.	Fish store
9.	Small food store	23.	Supermarket
10.	Carpenter	24.	Fish store
11.	Fishing Cooperative	25.	Wine store
12.	Store for coal	26.	Post-office (GPO)
13.	Rice and wine store	27.	Ryōkan (Japanese inn)
14.	Plumber	28.	Store for women's and children's clothes
15.	Store for electric equipment	29.	Fish store
16.	Ikebana school	30.	Tobacco store
17.	Taxi	31.	Sweet shop & stationer
18.	Fruit & vegetable store	32.	Wine store
19.	Acupuncture	33.	Printing company
20.	Taguchi Kōgyō Factory	34.	Ryōkan

Map 7 Service Institutions, Shops and Manufacturers in Shingū.

in Nakamachi and a carpenter in Honmachi which both employ local people as does the biggest shop, a rather small supermarket. The remaining shops employ only family members. A few are employed at the post-office and at the fishing cooperative (see map 7). Four women peddled fish and a couple peddled vegetable. Most of the people in Shingū work outside the ward, however, and Table 1 gives the place of work for 193 household heads in Shingu-kū.

Table 1 The place of work for 193 household heads calculated in per cent.

work in / live in	Honmachi	Nakamachi	Uramachi	Shingū
Shingū-ku	39.6	31.5	15.1	28.0
Shingū-machi	17.5	26.3	30.1	24.9
Fukuoka City	33.3	28.1	42.5	35.2
Kitakyūshū City	1.6	–	1.4	1.0
Other	11.1	17.5	16.4	15.0
Total	103.1	103.4	105.5	104.1

Notes: The total sums exceed 100% as some have work in several areas.

All the fishermen have been regarded as working in Shingū-ku though they go outside its borders.

Most of those working at "other places" have their place of work in Koga-chō (at least 7.3%) and in Fukuma-chō (at least 2.1%).

The composition of the working-force has steadily been changing from primary to secondary and tertiary industries. This tendency is particularly strong among the farmers in Shingū-machi while the fishing population has declined to a lesser degree, at least during the last twenty years from which we have information in the form of official statistics (Table 2). These numbers do not only include the Shingū fishermen, but those on Ainoshima as well.

There are about forty persons in Shingū whom I will regard as fishermen, and they are distributed in thirty households. By a fishing household I will mean a household where at least one member (usually the head) has fishing as the main occupation. Fishing does not need to be the most important source of income for the household. In several of the households which I have defined as fishing households, there are members who earn substantially more than the members employed in fishing. A fisherman is then a person who has fishing as the main occupation, though the income can be small.

It is difficult to trace the decline in the number of fishermen as the data are inadequate and conflicting, and different categories of fishermen are used in different documents kept at the fishing cooperative. A member of the cooperative was sometimes regarded as the household unit, sometimes the shareholder and sometimes the active fisherman, usually without stating which of the categories being used. A tentative table depicting the trends can nevertheless be made where the figures are arranged in four columns: (1) the household as member of the cooperative, (2) the shareholder as member, (3) the active fisherman regarded as member; and (4) the

Table 2 Development in the Occupational Structure in Shingū-machi
(Sources: Fukuoka-ken 1955, 1960, 1965, 1970, 1975)

Occupation	1955		1960		1965		1970		1975	
	House-holds	pers. in household	Men	Women	Men	Women	Men	Women	Men	Women
Agriculture	446	2971	647	783	547	610	459	505	377	417
forestry, hunting	8	43	3	1	6	1	7	1	4	1
fishing	164	1005	298	29	229	3	214	7	200	9
mining	4	18	17	0	5	1	2	0	2	1
construction	89	488	184	50	240	64	267	42	418	94
factory-work	110	608	277	123	479	301	631	492	726	464
sale, wholesale	96	471	194	231	241	269	404	446	548	485
bank, insurance	15	93	21	24	32	44	46	57	33	60
real estate agent	–	–	–	–	–	–	–	–	18	3
communication	128	628	189	15	269	37	644	62	717	56
gas, water, elec.	–	–	24	2	16	1	14	5	29	2
service	143	683	260	260	320	304	407	397	451	473
officials	49	287	82	25	69	19	98	51	115	40
other	153	665	1	1	1	0	1	2	3	16
Total	1405	7960	2197	1544	2454	1654	3194	2067	3641	2121

maximum number of fishermen engaged in any kind of fisheries as reported in the annual reports of the cooperative. If we assume that all the fishermen participated at least in one kind of fisheries (an assumption that will be justified in Chapter V), these figures will also indicate the scope of the fisheries in terms of man-power (Table 3).

Table 3 Persons and households engaged in fishing in Shingū between 1916 and 1935

Year	1 household as member	2 shareholder as member	3 active (?) as member	4 maximum number of fishermen in any fisheries
1916		78		241
1919	72		130	
1921			180	220
1922	62		150	
1928	59	79	146	160
1930	62		105	160
1935	50			120

The tendency is clear: all the columns (except column 2) show a steady decrease of the fishing population, but a few comments on the actual size of the fishing population are needed. The number of households engaged in fishing was probably somewhat higher than indicated in column 1 since several households were not members of the cooperative. The discrepancy between columns 3 and 4 is striking, but is at least partly explained as column 4 included in addition to the ordinary fishermen of Shingū also farmers who were seasonally employed, and probably the fishermen's wives who participated in pulling large beach seines used in the sardine and sea-bream fisheries. It is worth notice that the figures in column 4 show a sharper decline than those in column 3, indicating that the decrease in farmers being temporarily employed in fisheries went faster than among the fishermen proper. This is what we could expect considering the improved opportunities for employment in the industrial sector and the development of the agriculture with double-cropping reducing the off-season to practically nothing.

To summarize: the number of fishermen in Shingū started to decline early in this century, and by 1935 there were probably about one hundred active fishermen in the community distributed in about sixty or seventy households. The size of the fishing population remained on this level (except for a temporary decline during the war years) till 1950 when the number of fishermen again began to decrease.

Shingū-ura was founded on a dual economy; fishing and trading. The trading was to a great extent dependent on the fisheries, but the village was also a centre for the farming villages in the vicinity. This dualism in the economy gradually disappeared as the infra-structure changed, and most of the people are today working as employees in companies outside Shingū. The fisheries have declined in importance, and many people born in the community know neither the technology the fishermen use, nor the kind of fish they catch. The fishermen are almost an isolated community within the community. Even the fish-processing industry is independent of their products; formerly they had a decisive influence on their lives.

Four Merchant Houses

Four merchant houses (three in Shingū and one in Minato) played dominant roles though minor merchants also might have played important roles in the fishing business. To understand the structure of the fishing industry in Shingū in the past, it is necessary to introduce these four houses and ask: which assets did they possess?

The most important merchant house during most of the Tokugawa period was Mantokuya (later called Kaneuchiya), operated by the Kaneuchi family. People still vividly imagine this family, though it was at its zenith about 200 years ago. One of my informants had read (but could not unfortunately remember where) that Kaneuchi was one of the five richest men in Chikuzen; another told that Kaneuchi could walk all over Kasuya County without stepping on other men's ground; a third said that he owned ships and traded with Ōsaka, Kyōto, Korea and China; a fourth told me how his great grandfather, about 150 years ago, was forced to move from Nakamachi where Kaneuchi needed the space for his workers. "Kaneuchi decided everything". None could, however, date the decline of this family, and most of them guessed about a hundred years ago and connected it with the Meiji Restoration.

The oldest document found so far referring to the family is the temple book, stating that Kōjiya Shinzaemon died in 1677.[16] It was during the next generations that the house emerged as a powerful local force. The grandson (?) Kichisaburō became village headman in 1718, took the name Shinzaemon eight years later, and emerged as the district headman in 1734, a position he kept until he retired in favour of the eighteen-year-old son (adopted?) Shinzaemon Tokusada Tetsuroku in 1753.[17] During this time Mantokuya emerged as the largest land-owner in Shingū and Minato; in the Kyōho era alone (1716-1736) the house took over 133 fields totalling five *chō* or about fifty hectares.[18]

It is difficult to say when the decline started, but the loss of a ship loaded with 300 *hyō* rice in the 1780s was probably a serious blow, and Mantokuya borrowed 15 *kan* silver from the authorities. The absence of Korean missions after 1764 might also have been a cause for the decline, and Kitamura states three other causes as well (1976:12-13): (1) different taxes such as *goyōgin* (forced loans levied by the lord), rice to poor villagers etc., (2) the income from the land hired out to cultivators was low and gave an annual income of only five per cent on the investment, and (3) the investment in the fisheries did not seem to have been profitable.

By 1801 Mantokuya was in serious trouble, and they were given a loan of not less than 150 *kan* silver for which they had to mortage farming land, their house and nine storehouses *(kura)*. In 1808 their *sake* brewery and the pawnshop failed (Yayama Doc.). In 1822 Shinzaemon Kichijurō, the third son of Tokusada

Ill.8 Kaneuchi Kichisaburo. He was the first Kaneuchi to become district headman, and he donated most of the money for the rebuilding of the temple in the 1740s. The painting was donated to the temple in connection with the fiftieth anniversary of his death.

Ill.9 The remains of the once impressive house of Kaneuchi. The house was torn down in September 1978.

Tetsuroku, retired as district headman, a position that was given to Akaishi in Hakozaki-ura. The decline continued, and finally Kaneuchi Chūsaburō wrote an appeal to the authorities (*gun wakiyakushō*) in order to get back some of the land the house had hired out to other people. In his letter, dated the ninth month of Keiō 4 (1868) Chūsaburō stated that the house declined during the time of his father Kichisaemon (died 1867), and that they sold land, rented houses, sardine net and even furniture to pay huge loans his great uncle Junsaemon and Shinzaemon had raised. This was apparently not sufficient since land and house had been confiscated by the authorities and Junsaemon was exiled to Ōshima (Kaneuchi Doc.150).

Despite all these difficulties, Kaneuchiya remained one of the important houses throughout the Tokugawa period. Several generations served as guardian for children (*yōiku-kata*) in the district, Kichisaemon became village headman in 1843 and Shinshichi alias Chūsaburō became headman in 1874.

The end came, however, soon afterwards when Kaneuchi Sadao tried to establish a business in Hakata. This attempt failed and even their house was sold.[19] Sadao's son, the last male Kaneuchi, died in 1950 only thirty-five years old.

The Kaneuchi family does not exist any more, but their memory will live long in Shingū. Everybody "knows" that Shingū-ura was moved and founded by them, as was the temple.[20] Many of the inhabitants of today pass daily under a *torii* (Shrine-gate) donated by Shinzaemon Tokusada Tetsuroku, and their old house dominated the western end of Uramachi till it was torn down in September 1978 to give space for apartments.

When Kaneuchi declined, the house of Yayama rose to power. The descendants claim that their ancestors originally were among the "masterless" warriors (*rōnin*) who were reduced in status at the beginning of the Tokugawa period, a very common background for merchants and rural leaders (Sheldon 1958). (Kaneuchi was said to be of the same background.)

Yayama is also an old family in Shingū, and in the old temple their memorial tablets were placed on the most honoured place next only to Kaneuchi. The first entry in the temple books is from 1743, and the house was known as Kōjiya, which was located in Nakamachi. (Incidentally Mantokuya also started as *kōjiya.)*

Only a few documents are left from this family, but one scroll tells us that the family head was allowed to wear a sword in 1826 and allowed to take a surname four years later. Both the swords and the surnames were the privileges of the warrior class, but influential commoners were sometimes bestowed these privileges, usually in return for a high fee. In 1835 Yayama Hosuke received the same social rank as the district headman (*ōjōyakaku*) and four years later became village headman, a post he resigned in 1843. He managed to marry his daughter to a warrior, another way to obtain prestige, and in this connection he let his portrait be painted by the highest ranking painter in the province.

The basis for the economy of the house seems to have been, like Mantokuya, the *sake* brewing. This house went into a very abrupt decline, however, as Yayama Shōbei's father died in 1863 and his grandfather the next year. Shōbei was only a boy, the business declined and the *sake* brewery went bankrupt. Shōbei started his career as an interior carpenter, but failed. This time he lost the house in Nakamachi, and moved to a friend where he lived till his death. The descendants (a widow of Shōbei's grandson and her son) live in Fukuoka City.

The Yayama family left only one inscription in Shingū; on the wall which surrounds the shrine compound. But this inscription is something of a puzzle. On the wall it is engraved that *daihochō* Yayama Yauemon Yasuda donated this in 1843. (*Daihochō* is an alternative title for *ōjōya* or district headman.) Who was this man? The temple books mention "Yauemon" several times, but of little help. Just as many of the Kaneuchis were called "Shinzaemon", so were the Yayamas called "Yauemon".

The third house I want to introduce is Toyogaya, owned by the Toyomura family. This family is as ancient as the others; I was told that the village headmen in Minato-mura – Shinozaki Zenbei (died 1601) and Shinozaki Chōsaemon (died 1642) – were ancestors of this family. Not much is known about this family during the Tokugawa period, but it has probably been among the more influential. The family had allegedly some connection with the house Sakanaya (lit. fish shop) and Kitazaki Heishirō, who was village headman when he died in 1742. Kyūuemon, who was village headman in 1812, was again headman in 1931 when he changed his name from Toyogaya Kyūuemon to Heishirō.

During the last years of the Tokugawa period this house emerged as the leading one in Shingū, and it seems that they took over Yayama's *sake* brewery and buildings. Towards the end of the nineteenth century Toyomura had become the biggest property owner in Shingū. Toyomura Heishirō owned in 1888 thirteen compounds in Shingū, six of them farming lands, totalling about 4000 square metres. But this house also disappeared, and they allegedly sold much property to the leading house in Minato (see below). A couple of years ago they sold their last piece of land to Shingū-ku.

Yorosuya is strictly speaking not from Shingū, but as this house was very important to the fishermen long into this century, I find it correct to include it here. Unlike the three others, Yorosuya was based on a farming family in Minato-mura. In this village there were several strong households, each probably a main house (*honke,* cf. Chapter IV) for a group of households. One of these took the name Sakai. Chūkichi was the head of the main house, and he had three sons. The oldest took over the farm; the second – Shichiuemon – established a branch house (*bunke*) and did some farming and peddling to Hakata; the third – Sōhei – was adopted by a sick old woman who owned a debt-ridden *sake* brewery in Hakata. Both Sōhei, his wife and two children were adopted. Sōhei proved himself an excellent businessman; he put the *sake* brewery in good shape and bought several others. He became one of the very richest men in the area. "He carried the economy of Hakata on his back", according to the newspaper *Kyūshū Nippō Shinbun* (1935, vol.1-13).

Sōhei helped his brother Shichiuemon to establish a *sake* brewery in Minato-mura, and this prospered. By 1896 Sakai Manshichi (Shichiuemon's son) was the second biggest landowner in Kasuya County (Kasuya-gun 1924). He was the only son, but a husband was adopted into the family to marry Shichiuemon's daughter in order to establish a branch household. The branch house, called Shintaku, started production of oil and soy sauce. In the next generation Shintaku likewise adopted a husband for the daughter in order to establish their branch; this household set up an oil-shop in Uramachi. I will discuss the family system in more detail in the next chapter, but I will here mention that adoptions of husbands for daughters have been a typical facet of the "Japanese family system". This has been done to secure heirs, not to establish branch households. The rich families could, nonetheless, adopt husbands to convert economical wealth into a big and powerful structure and Sakai is a typical case. (Cf. Nakane 1967:169 who writes that this practice is "rather an exceptional case".)

The whole complex was by most people known as Yorosuya which also opened a shop for clothes, food, wood etc., and older people in Shingū can still recall this shop. Yorosuya, as well as Shintaku, both operated as banks and pawnshops, and land was frequently put as guarantee. One informant indicated that Yorosuya had enough land to harvest 4000 *hyō* and Shintaku enough for 1000 *hyō* rice.

But also Sakai reached its zenith. First to run into trouble was Shintaku, due to the introduction of electricity which was a blow to the production of oil. The third generation is now a tailor in Shimonofu. The branch of Shintaku moved out from Shingū-kū some years ago when their shop burnt, while Yorosuya's *sake* production is still going on in Minato under the brand name *Haretsuru*. The house did, however, lose most of its land during the land-reform introduced by the American occupational forces.

I have tried to sketch the history of the four most important houses in Shingū from about 1700 until today. There are several differences between them, but also striking similarities. The opportunities open for investment were limited (Hirschmeier & Yui 1975:34), and it seems that all four houses based their economic power on the same five pillars: land-holding, banking/pawnbrokerage, trade, *sake* brewing and ownership of fishing nets. It is well documented that Kaneuchi, Toyomura and Sakai operated as net-owners, and there is strong evidence that Yayama did the same. The merchants during the Tokugawa period were important sources of capital for the fisheries. Furthermore, Kaneuchi, Yayama and Toyomura all held important positions both inside the village and in the district, and thus had judicial authority over the estate of the village, among which the fishing grounds were the most important. Sakai, on the other hand, lacked this political power over Shingū.

One more point should be mentioned: during the winter when the weather conditions often prohibited fishing, the fishermen could find employment in the *sake* breweries. The fishermen of Shingū were working in this industry till World War II. There were thus strong mutual interests between the merchants and fishermen.

We are now ready to proceed to an investigation of the history of the fishing technology, but first it will be useful to introduce some more ethnographic facts on the social organization of Shingū today, and this subject will now be further considered.

CHAPTER IV

HOUSEHOLD AND COMMUNITY

The basic unit of social organization in Japan is the household. The village is usually conceived by the inhabitants as a group of households, and the size of the community is often given in number of households rather than by number of individuals. The household is integrated into the community through several institutions; kinship groups, localized groups, occupational groups, and religious associations. Despite this, there are tendencies that the fishermen are increasingly isolated from the rest of the population in Shingū, but in order to understand this process it is necessary to analyse the framework in which the household operates.

The Household

The household *(ie)* or domestic group is more than a family; it is conceived as a permanent institution where the actual member "at any one time is only the current concrete manifestation" (Fukutake 1972:39). The household is a corporate group where the present members are merely in charge of the household and its property which includes the house, fields, tools, fishing-rights, tombs, memorial tablets and certain social relations. The person "in charge" of the household – the head of the house – has received his position from the ancestors, and he is expected to transfer it to one of his children. While in office it is his duty to see to the welfare of its members, and if possible to increase its wealth and improve its social standing.

The househead's position in the household is closely linked with the concept of *on*, which is an obligation incurred by a received favour. The greatest *on* one owes to one's parents (formerly allegedly also to the emperor and lord). As the children owe *on* to their parents, so the head owes *on* to his, i.e. the ancestors. The obligation to repay the debts is not terminated by death, and the living members of the household have certain ritual duties towards the household ancestors. Some of these rituals are performed daily in the front of the household Buddhist altar *butsudan;* some rituals are performed on fixed days like New Year, equinoxes (*higan*) and the festival for the souls (*Bon*) in August; and others again are performed on occasions like memorial days (*hōji*). Failure to perform these rituals can incur anger and disaster inaugurated by the ancestors, but also sanctions from relatives and gossip from fellow villagers.

Thus the household can best be viewed as a permanent structure where the living personnel changes, but where membership is not ended by death. To phrase it with Nakane (1967:2): "(*Ie*) is conceptualized in the time continuum from past to future, including not only the actual residential members but also dead members, with some projection also towards those yet unborn."

As the household is a corporate group and the *on* to one's parents can never be repaid in full, it becomes important to continue the household as a unit in perpetuity. It is then of the utmost importance to find an heir. Ideally the succession of the headship goes to the eldest son, and this was even legally sanctioned by the old civil code abandoned after World War II.[1] If there was no able son to inherit the household, it has been common to adopt a son-in-law to one of the daughters, preferably the eldest. One fishing household in Shingū was forced to adopt son-in-laws in three successive generations.

Whether the successor is a son or a son-in-law, he traditionally inherits the household property and occupation (Befu 1963:1330), and there is little room for other children to remain in their natal home after marriage. Excess daughters are married out, while several possibilities are open to the sons. Out-adoption was a likely solution in the past when there were few other possibilities, but since the industrialization started, it has been easier to establish new households both in the community and in urban centres. Younger sons have often been provided with better education than the eldest for in this way to make a better start. Great social mobility has occurred; one fisherman's son is at present one of the directors in one of the largest concerns in Japan.

Frequently the eldest son did not inherit the house, but moved out. One reason could be personal desires or parental preferences. (One case where the eldest son was disinherited is mentioned in Chapter VII.) Sometimes the parents move to a separate house upon marriage of the eldest son, leaving him and his bride in the old building. The parents bring younger children with them, however. The new house created in this way has been termed *inkyo* house, or retirement house (cf. Chang 1970). This institution sometimes caused disputes over property and the legality to the headship. (See Chapter VII for a case illustrating this point). So frequent have such disputes been that the authorities in Tokugawa Chikuzen found it necessary to issue a decree in 1772 stating that the people did not follow an ideal pattern. Many fishermen moved with their younger children upon the marriages of their first sons, and gave the property to the former. This caused grave resentment, and the authorities suggested that the older and younger brothers should work together. When the younger brother obtained a wife, they did not need to employ a maid, and the household would grow richer. Only later should it be possible to separate the main and branch households (Fukuoka-ken 1963:655).

The household consists today ideally of a stem family, i.e. a married couple, their unmarried children and the parents and unmarried siblings of one of the spouses, usually of the husband. Occasionally more distantly related persons are living in the house, and if servants are engaged, these are included in the household as well. This was, however, more common in the past, and at present the fishing households contain no servants.

Of the twenty-nine households engaged in fishing in 1976, three contained only one generation (one divorced man, one newly married couple, and the third a couple without children married a long time), sixteen households had two generations and ten had three (Table 4).

Of the households with only two generations six were established by the present head, and these will probably develop into the ideal stem family. The other ten households with two generations are in this situation due to their stages in the developmental cycle, and their situation is temporary. Among the fishermen in Shingū the ideal of the stem family is thus still strong.

There are no indications that there have been any marked changes in the structure of the household since the Tokugawa period. As we have seen, the authorities

Table 4 The composition of the fishing households in 1976

No. of generations in the household	No. of persons in the household 1	2	3	4	5	6	7	8	Total
1	1	2	–	–	–	–	–	–	3
2	–	1	7	7	1	–	–	–	16
3	–	–	–	–	2	6	1	1	10

issued decrees aimed to encourage extended families and delay fission. The decrees did probably have no effect. The mean household size (MHS) has been rather low during the whole period under discussion. We do not unfortunately have any documents from the Tokugawa period giving both the size of the population and the number of households for Shingū-ura. The number of households seems to have been extremely constant; there were 165 houses in 1692 including twenty-three empty ones, and in 1889 there were 165 households. A map of the village from about 1820 indicates approximately 170 households. The population on the other hand was 666 in 1825. This implies a MHS of four or less, which does not disagree with estimates made by Nakane (1974) who gives a figure for Tokugawa Japan of five or less, and R.J. Smith (1974a) who gives a MHS of less than three for some groups.

The low MHS for the Bunsei era (1818-1830) indicates either establishment of new households through fission or a low fertility. As the number of households seems to have been fairly constant throughout the Tokugawa period since the relocation of Shingū in 1685, few new households have been created. The fertility must have been low as out-migration was probably rare due to prohibition.

A striking feature of the population of that time was the difference in size between the male and female populations.[3] What can be the reason for this big difference? One suggestion could be that Shingū attracted a large workforce from

Table 5 The population of Shingū in 1825.[2]

	Men	Women	Total	Ratio
11 years or older	303	243	546	125/100
10 years or younger	67	49	116	136/100
Yamabushi (hermits)	2	0	2	–
Absent from Shingū	2	0	2	–
Total population	374	292	666	128/100
Babies	8	5	13	160/100
Deaths last year	10	5	15	200/100

outside. There were many merchants who could attract apprentices; a common feature during this period. Time and again, moreover, farmers were forced to leave their land because of poverty, and there might have been opportunities for some work and for begging in Shingū.[4] This cannot, however, explain the difference in the numbers of male and female children.

Though infanticide was forbidden, it has been widely practised in Tokugawa Japan. Smith (1977) found that infanticide was practised in order to plan the size of the family and the balance of sexes. Due to the Japanese family structure stressing the importance of male heirs, a surplus of boys is to be expected from the practise. If a male heir was already secured, forced abortion or self-imposed restraint may have been practised, which also leaves a surplus of boys. If this is so, the unbalanced sex ratio among the children will be particularly pronounced in the small families. Smith (op.cit.:76) thus found that "a clear relation emerges between completed family size and the sex ratio of children. The larger the family, generally speaking, the more evenly balanced the sexes; the smaller the family, the higher the proportion of boys."

Table 6 MHS of Shingū-mura and Shingū-machi

	Shingū-mura							Shingū-machi	
	1868	1889	1916	1926	1935	1950	1960	1970	1978
Pop.	2508	3105	3927	4371	4237	5618	8420	10824	13340
H.holds	495	551	636	697	720	1019	1676	2649	4025
MHS	5.07	5.64	6.17	6.27	5.88	5.51	5.02	4.09	3.31

The ecological adaptation also supported such a selection. Among the farmers several sons could be a burden, and a threat to the farm as a viable unit. Commonly, division of estates yielding less than ten *koku* was prohibited. The daughters, on the other hand, were married off, and before their marriage they contributed importantly to the labour force both in the house and the fields. Among the fishermen the situation was somewhat different. Two or more sons could be an asset to a fisherman as they could directly participate in the fisheries. In large net groups the households were paid according to the number of members they sent to the undertaking. In Kujūkurihama district "even a family that had an heir would receive a boy into the family as a son to be sent to a fishing group as a fisherman" (Kujukuri Research Committee 1958:29). The daughters could not participate in the fisheries in the same way, as they were barred from such participation due to religious beliefs (cf. below).

Towards the end of the Tokugawa period and during the Meiji and Taishō eras (1868-1926) the MHS increased from 5.07 in 1868 for Shingū-mura to 6.27 in 1926 (Table 6).

This development has been typical for Japan as a whole, though the MHS for Shingū-mura was somewhat higher than for the country. The MHS is not uniform throughout Shingū-machi, however (Table 7).

Table 7 Population in Shingū-machi by ōaza 30-4-1978

ōaza	households	population	MHS
Matano	33	133	4.03
Tachibana-guchi	138	648	4.70
Harugami	193	678	3.51
Mishiro	122	494	4.05
Kaminofu	878	2903	3.31
Shimonofu	2127	6421	3.02
Shingū	252	980	3.89
Ainoshima	198	702	3.55
Minato	84	381	4.54
Shingū-machi	4025	13,340	3.31

There are great variations in the MHS between the *ōaza;* the farming *ōaza* have all MHS of more than four, while the others have less than that. There are thus obvious correlations between occupation and MHS. In 1955 the MHS of the farming households in Shingū-machi was 6.66, that for fishing households was 6.13 as compared to 5.63 for the whole township.[5] The MHS for the fishing households in Shingū in 1975 was 4.21 as compared to 3.80 for Shingū.

The changing MHS does not reflect changes in the kinship system; rather it reflects the changing fertility of the population. The fishing household has throughout the period under discussion typically consisted of a stem family; only seldom did two married siblings live under the same roof. The consequences of the rather small MHS during most of the history of Shingū will be fully analysed in Chapter IX; here it will suffice to mention that households were often forced to cooperate in the fisheries (cf. particularly Chapter VII). This cooperation between households was often based on kinship relations, and this subject we shall now examine more closely.

Dōzoku and Shinrui

The relatives in Japan are conceptualized within two institutions: (1) a group of households based ideally on patrilinear descent (*dōzoku*) and (2) the bilateral and affinal kin (*shinseki* or *shinrui*).

No aspect of the Japanese social structure has received as much attention as the *dōzoku* group, and the views are highly contradictory. It has been defined as a patrilinear lineage (Befu 1963, Fukutake 1972, among others), as a cognatic descent group (Brown 1966) and as not being a descent group at all (Nakane 1967). Most writers agree that the *dōzoku* is a vertical structure with a main or stem household (*honke*) at the apex and branch households (*bunke*) under its authority. Usually the branch households are directly linked by consanguinal ties (agnatic) to its main household, but this is not always the case. Frequently an heir is an adopted son or son-in-law, and in the past it was not uncommon to put up unrelated tenants as branches. This fact leads Nakane to the conclusion that "the *dōzoku* is *not* a patrilinear lineage" (1967:85). And she continues: "The *dōzoku*, according to my definition, is a set of households which recognize their relationship in terms of *honke* and *bunke*, and which on the basis of this relationship have developed a corporate function as a group." (op.cit.:90:91). The *dōzoku* is thus a "group of *households*, not of individuals who are classified by patrilinear descent" (Ibid. 84). Nonetheless, it has been rare since the end of the Tokugawa period that unrelated households were established as branches, and Fukutake concludes that *dōzoku* "tended to be exclusively kinship groups" (1972:71).

Usually it is second and succeeding sons who are established as branch households, but occasionally daughters are put up in independent branch households, and Brown views this as an indication that the *dōzoku* is a cognatic group. However, if such a measure is taken, it has to be justified in some way. Brown himself writes:

> (Cases where daughters establish branch households) are frequently found in response to definite problems. One such problem is what to do with a daughter who is weak, sickly, deaf, deformed, or just generally unattractive. Her family will certainly have a difficult time in arranging a marriage for her. They therefore may have to take serious measures to relieve themselves of her support and to enable her to lead a somewhat normal married life. They often can entice a young man, most likely a junior son from a poor family, to marry her as an adopted husband if a house and some land are promised the new couple (Brown 1966:1133).

Such cases are known in Shingū, and daughters may also be established in branch households if the *honke* is prosperous and wants to increase its field of influence (cf. Yorosuya in Chapter III). But these are exceptional cases and explanations for such measures are required by the villagers. On the contrary, if a son is put up as a branch, this is regarded as the most natural thing to do, and there will be no further questions as to why this happened. Today then, for most analytical purposes, the *dōzoku* is a vertical structured group of households where the househeads are ideally patrilineally related and where a household is a member of this corporate group only through an active *honke-bunke* relation.

The strength of the *dōzoku* is not uniform throughout Japan. In its "purest" form it is only found in isolated mountain areas in the Central and Northeastern parts of Honshū. In many places it is hardly found at all. The presence or absence of *dōzoku* has often been analysed in terms of evolution. It has been assumed that this kinship group has been universal, but due to economic developments during the Tokugawa period and after, it declined (Smith 1959, Cornell 1964). This should allegedly explain why *dōzoku* is found in the mountainous Central and Northeastern Honshū. Johnson (1976) who did field work in Nagura-mura – an isolated village in the above-mentioned mountainous area but where *dōzoku* does not exist and probably never did – argues convincingly against this view, and he turns to historical explanations. In particular he attributes the presence of *dōzoku* to rural ex-warriors and to merchants, and referring to works by Beardsley (1951) and Kujukuri Research Committee (1958) he concludes that the development of a *dōzoku* depends on a set of special conditions:

> These conditions would be either (ex-warriors) with large holdings and strong local power, or urban merchants investing in rural areas and controlling the dōzoku through their access to non-agricultural wealth (Johnson 1976:214).

In Shingū there were both ex-warriors and merchants, but strong *dōzoku* does not seem to have developed. It will be necessary to consider the ecological conditions as well as cultural determinants in our analysis (Southall 1971:161). In doing so it is useful to distinguish two aspects or levels of a *dōzoku;* the economic and the ideologic. Our task is then to describe the processes that form and maintain each level, and to investigate the interrelation between them. This requires a separate paper, but some comments will suffice at this juncture.

By the Tokugawa period it seems that the ideology of descent that continued right up to the end of World War II was firmly established, at least among the land-owning *élite.* During the Tokugawa period this ideology was strengthened by at least two factors. First, the edict forcing everybody to register at a Buddhist temple brought the whole populace into contact with the Buddhist form of ancestor worship (Yanagita 1970:94). Secondly, the authorities put great stress on the neo-Confucian world-view, and Confucianism filtered down through the population so hardly any was unaffected. The basis for this conscious model was the concept of filial piety (*kō*) which was one of the important elements in Confucianism in Japan, and by the end of the eighth century the *Classic of Filial Piety* was taught in all schools, "and every child who knew how to read could recite it by heart" (Bellah 1970:89). Filial piety was (together with loyalty) the highest virtue, and it was closely connected with the concept *on.* Together they created an absolute obedience towards the parents, and reverence towards the ancestors. It was not only the duty of the descendants to take care of the welfare of the ancestors, but it became the exclusive right of the child that perpetuated the household. By the Muromachi period (1392-1568) primogeniture seems to have been institutionalized (Sansom 1946:365), and other children who wanted to pay respect to the souls of their parents, grandparents etc., could do this only in front of the Buddhist altar at the main house. As soon as the founder of a branch household died, this house would get its own altar and its own ancestor to worship, and the household grew less dependent on the main house for religious services.

Here we have the main factors that form the ideology of a *dōzoku;* on the one hand the concepts of filial piety, *on* and ancestor worship, and on the other hand the patrilineal bias and the exclusive rights of the heir to keep the memorial tablets of the ancestors. This ideology received strong support throughout the Tokugawa period, and even in the present Civil Code, article 897 reads: "The ownership of genealogical records, of utensils of religious rites, and of tombs and burial grounds is succeeded to the person who is, according to custom, to hold as a president the worship of the memory of the ancestors" (Minister of Justice 1966:521 cited in Smith 1974b:34).

The model outlined above is followed by the great majority of the Japanese. Throughout the country the terms *honke* and *bunke* are used and understood, but this is not sufficient to form the typical *dōzoku* as reported by so many writers, and now we have to look at the other level, i.e. the ecological factors that make possible the economic structure found in the strong *dōzoku.* Nakane who draws on rich sources put forward two prerequisites for formation of *dōzoku* (1967: 119-120):

First, the community needs to provide substantial resources for the maintenance or the accumulation of wealth by one sector of the community. And it should be difficult for village members to obtain a living other than by dependence upon the limited resources within the community.

and

Second, the existence of effective dōzoku seems to be largely determined by the earlier pattern of settlement of the village, and a high degree of stability in the economic organisation of the village.

Shingū-ura was, as we have seen, a commercial centre where the economy was rather unstable. This holds true both for the merchants and the fishermen. The fishermen had difficulties in accumulating wealth, and one of the reasons is the effect of inheritance. Creation of a branch household would imply division of property. For a farmer this would mean a permanent loss of land. Where a branch household was established, this usually received a small share of the land, often too small to live on. In this way the branch household remained dependent on its main household and often became the latter's tenant. Thus, the agricultural adaptation could well be a base for development of *dōzoku*. For the fishermen, on the other hand, division of property might only mean a temporary loss. Neither boats nor nets are durables as is the land. Wealth was therefore difficult to accumulate within the traditional fisheries. Success as a fisherman could imply a faster division of property, as a son could be helped to establish a branch household and he would receive the old boat. Occasionally the eldest son might have been established in branch households as the younger sons were still too young to benefit from a division. When a branch household was established in this way, it became difficult to keep this in dependency. A fisherman's success depends on skill and luck, not on descent. Furthermore, since the Meiji Restoration Shingū has been brought into the economy of a fast developing city like Fukuoka with its opportunities for employment. The basis for an effective *dōzoku* is therefore not present.

There are hardly any kinship groups in Shingū today that can be called *dōzoku*. The concepts *honke* (or more commonly the synonym *honya*) and *bunke* are used, but *honke* refers usually to the natal house of the household head only, and *bunke* to households established by brothers who have moved out of the parental house. The relations between such households are intimate as long as they are living in proximity. They do not form corporate groups in economic activities, but the *honke* does have some importance in the religious life of the fishermen since the ancestral memorial tablets are found there. As the ancestor usually means the founder of the household, his descendants who remained in the house and their spouses, a branch household will in its second generation have its own ancestral tablets, and the importance of the *honke* loses much of its meaning. Nonetheless, *bunke* used to visit occasionally their *honke* for ancestor rites, especially for New Year, *Bon* and the memorial services *hōji*.

Honke-bunke relations more remote than outlined above are sometimes recognized, but the relationship is usually of little importance unless strengthened by other institutions like the neighbourhood group. The insignificance of such relations is also illustrated by the shallow genealogies found in Shingū; only a few (all *honke* heads) could mention the names of their patrilineal great-grandfathers;

some did not know the names of their grandfathers. The pattern we then find in Shingū is very close to what we could expect when the conscious model that serves as the guide for the actors' behaviour is not supported by the economic structure.

Though *dōzoku* is not a typical feature of Shingū society, a few kinship groups resembling the *dōzoku* did nevertheless exist in the past, and they seem to have developed just when large management units were organized in the fisheries. In order to solve the organizational problems connected with large net-groups, one alternative was to turn to the ideology of *dōzoku.* The most important net-group organized along these lines in Shingū was the Kitagata-ami group (cf. Chapter VII). Similar groups existed in Fukuma-ura, Kōnominato-ura and Kanezaki-ura in Munakata County during the Tokugawa period (Tsukushi 1974:114).

The *dōzoku* is a corporate group, but our second concept – the *shinrui* – denotes only a category. Contrary to the case of *dōzoku,* a *shinrui* constitutes an ego-centred category of kin, and includes both bilateral kinsmen and affines with their household members. (See Nakane 1967:28 for a figure showing the usual extention of *shinrui.*) The *shinrui* does not then cover the same people as a kindred defined by Freeman (1961:201-202), a fact recognized both by Nakane (1967:27) and Befu (1963:1331). Befu makes the sound assumption that *dōzoku* and *shinrui* exist side by side, but that they have different functions. The *dōzoku* is "primarily concerned with transmission of family name, property and occupation" while *shinrui* is "concerned with the management of day-to-day activities" (op.cit.:1334). In Chapter VII we will, however, see that both *dōzoku* and *shinrui* relations were used as a basis for formation of net groups, and that they may be looked upon as alternatives.

Ill.10 The Bon Parade. During Bon the ancestors visit their descendants. Those who died during the last year receive special attention. These are sent back to Amida's Western Paradise in boats carried in a procession to the beach.

Local Groups

The households are not only integrated within the framework of kinship, but localized groups often assume some prominence. These groups fall mainly under three headings: (1) subdivision of the ward, (2) "age classes" for lack of a better term, and (3) occupational groups.

During the Tokugawa period the households were as mentioned in the previous chapter, organized into neighbourhood groups (*goningumi*) consisting of from five to twenty households. These were used by the authorities to maintain social control in the villages. Since that time the groups have received changing attention from the authorities. They were not recognized within the framework of the Meiji administration, but were revitalized throughout the country under the term *tonarigumi* during World War II in order to transmit "the will of those above to those below" (Fukutake 1972:97). After the war the groups were made illegal by the Occupation until 1952 when Japan regained her sovereignty. The neighbourhood groups exist today in most rural communities.

Despite the fluctuating emphasis the groups have received from the authorities since the Tokugawa period, they have always played a rather important role in Shingū. This means that the groups fill other functions than the administrative one, and some writers take this as an indication of the absence or existence of only very weak *dōzoku* (Befu 1963:1336).

The nineteen neighbourhood groups in Shingū-ku are headed by *kumichō* (group leader), and the position as *kumichō* rotates among the household heads of the group. If the household head is a woman, which only occurs in those cases where there is no adult male in the household, this household is by-passed until it again contains an adult male. The household heads do, on the other hand, often send their wives to represent the household at communal meetings (*sōkai*) or to meetings within the neighbourhood. The increasing participation of wives as deputies at such meetings reflects the declining importance of the local community for most of the inhabitants.

The functions of the neighbourhood group can be viewed as dual; it is a convenient link between the authorities in the town office and the households, and it fills social functions that some people still regard as important.

The duties of the *kumichō vis-à-vis* the administration are first of all to bring information from the town office to the members of their groups. This information goes either directly to the *kumichō* or more commonly via the ward headman (*kuchō*). The ward headman is one of three executives elected at the general community meeting in January, the other two being the vice-headman (*fuku-kuchō*) and the cashier (*kaikei*). These positions rotate between the streets in the following manner: if the headman is living in Uramachi, the vice-headman (and the next headman) should live in Honmachi, and the cashier (and thus the next vice-headman) in Nakamachi. It is nonetheless possible for a headman to remain in office for more than one period of two years, and commonly he does so as it is difficult to find willing candidates to this position. The work is rather demanding and time-consuming, and the salary was in 1976 only 232,000 yen as compared to 176,000 yen each for the deputy and cashier. This limits the candidates mainly to retired men. It is furthermore desired that the candidate has a

Ill.11 Honmachi seen from the sand dunes towards East.

wide network of social contacts, making a former civil serviceman an ideal headman. The headman from 1972 to 1976 was just such a man, but he was severely handicapped in his tasks as he was not born in Shingū but on Ainoshima. No other willing and suitable candidate from Uramachi existed, however. The new headman from 1976 was on the other hand a farmer living in Honmachi.

The three executives and the nineteen *kumichō* play a certain role in forming the policy of the ward and the township. They meet to make a request list presented to the town office, and to decide the budget of the ward. Usually the annual community meeting where all the households may be represented accepts the budget and the requests with few objections.[6]

Besides the formal tasks outlined above, the neighbourhood groups serve social needs. At present these groups perform miscellaneous functions, but there are great variations between the groups even along the same street. I interpret the lack of universality as an indication of social change. All the groups serve still as mutual assistance groups, particularly in the case of deaths and funerals, and usually some attention is given to marriages and births within the group as well, likewise to communal festivals. At least one annual party *(higomori)* in the group is held.

Though some of the women participate both at meetings in the neighbourhood groups as household heads or their deputies, and at the meetings with the ward executives as deputies for the *kumichō,* their husbands, the neighbourhood groups and the ward administration integrate the households in Shingū mostly through their male heads only. There are, however, associations that do integrate the households through other members than the heads. These associations have been labelled "age classes" (Norbeck 1953), and I will use the same term. Nonetheless, the actual membership in these associations is not determined by age alone, but also by the position the individual holds within the household. Few married persons participate in the youth clubs, and never so if the married couple live in a separate household. In that case the young wife will be a member of the Housewives' Association and the young husband will take his seat among the heads in the neighbourhood group.

The following age classes exist in Shingū at present:

(1) *Kodomokai* is an association for all the children in elementary and junior high school. The activities are mostly recreational. Though the officers are all children, the *kodomokai* is closely supervised by the school, PTA (Parent-Teacher Association) and *Ikuseikai* which is an association of parents with children in school.

(2) When the children graduated from school, they used to become members of the youth clubs, *Seinenkai* in the case of boys and *Shōjokai* in the case of girls. Both these clubs are defunct, however, and an explanation for this development will be provided at the end of this chapter. The functions of these clubs were, in addition to those stated in the previous chapter, to make preparations for festivals centred on the shrine. Today these tasks are taken care of by the *Shōbōdan* and *Kyūsaikai.*

(3) *Shōbōdan* or the fire brigade whose main task is of course to fight fires, is recruited by young men in Shingū, from about twenty to thirty-five years old. Besides actively fighting fires this association is also involved in fire-preventing activities. *Kyūsaikai* is the fishermen's fire brigade, but besides the fighting of fires, the main function is to rescue ships and people in danger on the sea; duties that are remnants of the *kako* institutions.

(4) *Fujinkai* is the association for the housewives, and its activities are many. One of the main aims is educational, i.e. to help the wives to improve the living conditions. During my first stay a campaign for new food habits, i.e. to eat more nourishing food, was under way, as well as a campaign against lavish spendings on weddings, births and the like. The association also runs courses in Japanese classical dance and kimono making, and it has even entered commercial activities. Through contacts with producers some bargaining deals have been made, and the members sell different items.

(5) *Wakabakai* is an association for middle-aged women who have handed over the running of the household to a young bride but who are too young to enter the Old People's Club. Of all the "age classes" this is the least significant in the daily activities in Shingū, but the association shows clearly the importance of the position held in the household.

(6) *Ryōjin-club* or the Old People's Club is open to everybody older than sixty-five years. This club is sub-divided into three, one for each street, since each division is large enough to obtain financial support from the township and the sum received is fixed regardless of the actual size of the club above the minimum. The activities are mostly recreational.

Finally, some words should be said about the occupational groups, of which Shingū has three; one for the businessmen *(Shōkōkai)*, one for the fishermen (*Gyogyō kyōdō kumiai*) and one for farmers (*Nōgyō kyōdō kumiai*).

Many of the fishermen are members both of the fishing and agricultural co-operatives. The agricultural cooperative is the oldest and soon developed many activities of which the banking business attracted the fishermen. Many had savings accounts in the cooperative bank. Some of the fishermen formerly owned a little farming land, and during the war years they were offered land cheaply by the government as a means to increase food production. The land was located near NH 3 in Kaminofu, but shortly after the war a vinegar company wanted to buy the land. The fishermen were in disagreement, but the majority of the executives of the fishing cooperative voted for sale, and since that time no fisherman has had any land in excess of a kitchen garden behind the house. Some are still members of the agricultural cooperative, though, and many use the cooperative store in Minato-ku.

The fishing cooperative is by far the most important association to the fishermen, and the activities of this cooperative will be discussed at some length in Chapter VIII. Here it will suffice to state that when the cooperative was founded, only fishermen were allowed entrance. This cooperative is thus exclusively for the fishermen, and the fishermen have tended to form a community within the community. This independence is further enhanced by the existence of the *Kyūsaikai,* which alienates the young fishermen from their age-mates in *Shōbōdan,* and by the cooperative's women section *Fujinbu,* with much the same functions as *Fujinkai.*

We have in this section surveyed the local associations within which the households operate. Religion has also traditionally been an integral force to reckon with, to which further consideration will now be given.

Ill.12 Ebisu-sama at Isozaki shrine. Ebisu-sama is one of the four deities that, together, are regarded as the tutelary deity (*ujigami*) of Shingū. Ebisu-sama is specially a patron for the merchants and fishermen.

Religious Affiliation

A stroll around Shingū-ku will reveal several religious objects. Most imposing are the Buddhist temple Sainen-ji, located in the western end of the ward, and the Shintō shrine Isozaki-jinja in the eastern end. There are further two small buildings in Uramachi where the bodhisattva Jizō is enshrined, one building dedicated to the deity Awashima, and several sacred stones. From some houses the prayer *namu myō hōren gekkyō* may be heard indicating that the households support the new Buddhist sect Sōkagakkai, and just outside Shingū-ku in Shimonofu one will find a church belonging to the Shintō sect, Konkōkyō. The complexity of the religious life is apparent, but space does not allow me to go into detail here. I will merely limit myself to the main outlines and to the religious beliefs directly affecting the fishing business.

Isozaki-jinja, though conceived as one shrine, actually consists of many shrines dedicated to different deities (*kami*); among them Ebisu, Inari, and Temmangū. But most important are the small Haenominato-jinja where Suminoe-no-mikoto (or Umi-no-mikoto), a deity for ocean and purification, is enshrined, and Isozaki-jinja proper where the three deities, Daikoku (the God of Riches), Ebisu (the patron god of both fishermen and merchants) and Susanoo-no-mikoto (often regarded as the ruler of the sea and tides) are enshrined. These four latter deities are collectively referred to as the tutelary deity or *ujigami,* and the *ujigami* of Shingū has thus a direct significance for the traditional industries of the community.

People who live in the ward and who regard the deity as their *ujigami* are called *ujiko.* In Honmachi as much as ninety per cent of the households regard themselves as *ujiko,* while only seventy per cent in Uramachi do so. This difference can be explained by the greater physical mobility in Uramachi. People who move only seldom continue to regard the deity of their former place as *ujigami;* on the other hand it takes time before they are completely incorporated into the new community. New settlers do not regard themselves as *ujiko* to any shrine (Table 8).[7] Likewise, people who move from Shingū cease to be *ujiko* to Isozaki-jinja. The *ujigami* is then a deity closely associated with a locality, and only those who identify themselves with this locality are *ujiko.*

The shrine in Shingū has a full-time priest (*gūji*) whose main religious function is to perform purification rites on behalf of individuals as well as the ward. He is assisted by six representatives (*ujiko sōdai*) elected by the *ujiko.* There are two representatives from each street, one each from the western and eastern section of the street. The period in office is three years, but it is common to re-elect the representatives as long as they wish. The position is solely an honourable one, and elder respected men who "know things" are chosen.

Central to Shintō is the concept *kami*, above translated as "deity". This is not accurate, however, and a better term would be "sacred power" which may exist in anything beyond the ordinary. "*Kami* may thus be described in certain people, in certain trees and stones, mountains and islands; in the excellence which overshadows the practice of certain crafts, in the continuity and protection which attends a family stemming from a remembered ancestor." (Blacker 1975:34). The *kami* "is neither good nor bad, but can manifest itself as benign or destructive

Table 8 Religious affiliation of 225 households in Shingū (answers to a questionnaire, December 1975)

When did your ancestors move into Shingū	Temple affiliation			Shrine affiliation		
	Ṣainen-ji	other	none/NA	Isozaki-jinja	other	none/NA
before 1945	127	23	8	140	2	16
1945-1960	7	5	1	8	0	5
after 1960	0	13	5	3	3	12
do not know/NA	8	12	16	16	2	18
Total	142	53	30	167	7	51

to human interests according to the treatment it receives" (ibid.:41). As long as the proper rites and conducts are performed, the *kami* will bring health and happiness to the worshippers, but if the *kami* is insulted, disaster may follow. Right treatment is devoted worship and offerings, but the worship has to be conducted in a state of purity. Nothing is more offensive to the *kami* than ritual pollution on the part of the worshipper, and death, child-birth and menstruation are particularly defiling. Taboos related to death, child-birth and menstruation are slowly dwindling in Shingū as in other places in Japan, but a few are still observed. It is regarded as offensive to the *kami* if a menstruating woman visits the shrine, and she is not allowed within the shrine precinct before on the day of *hiaki;* i.e. the day when the child is brought before the *ujigami* for the first time, on the thirty-first day for a boy and thirty-third day for a girl. Bereaved families are further supposed not to participate at any rites at the shrine during the period of mourning, usually forty-nine days. Within the household a menstruating woman does not perform the daily rituals in front of the god-shelf (*kamidana*), and the shelf is "closed" for a certain period after a death in the household. When the period is ended, the shelf is purified by sprinkling sand from the beach over it.

Thus death, birth and, to a certain extent, women are defiling and offensive to the *kami* whom people depend on to secure their livelihood. There is a common taboo against women participating on the fishing boats (Ōtō 1963, Schepers 1935: 182) though exceptions are found (Norbeck 1954, Segawa 1963:239). In Shingū this taboo has been strong, and only during the last few years have housewives started to participate on the fishing boats (cf.Chapter VI).[8] The defilement during menstruation is nonetheless still recognized, though the defiling period has been shortened and may today be regarded as the first day only. The women, though active fisherfolk, are on the other hand still not allowed to participate at the rituals during the annual festival *Ryūgū-matsuri* on 15th November, performed on some rocks called Isozakibana.[9]

Death is as already mentioned also polluting, and the bereaved family is affected by taboos. Nobody from the household is allowed to fish for seven days, but this period is extended, previously to thirty-five or forty-nine days, in the case of communal fishing like beach seining. Relatives are also affected, but for a much shorter period, and other fishermen are only affected as long as they are engaged in the preparation for the funeral.

This brings us to Buddhism that is usually referred to as a religion concerned with the death and the life thereafter. Being an *ujiko* does not in any way imply that the household does not belong to the congregation of a Buddhist temple. The two are not exclusive but rather complementary. "He looks to Buddhism as a means of soul salvation and to the pantheon of Shintō gods for protection in this life" (Embree 1939:221). Embree is not completely right as he also may look to the ancestors for protection, and the ancestor worship is today closely tied with Japanese Buddhism.

When the religious edict forcing all households to register at a Buddhist temple was enforced in the 1640s, most of the households in Shingū and Minato became parishoners of Sainen-ji.[10] Most of the households in Shingū and Minato have remained parishoners since, as have many in Shimonofu and in Mitoma in present Fukuoka City.

The organization of the congregation is in many ways parallel to that of the shrine. The priest is assisted by temple representatives (*tera sōdai* or *danka sōdai*) elected by the parishioners. Nakamachi and Uramachi elect two representatives each to the temple in the same way as to the shrine, while Honmachi elects three.[11] Finally, the parishioners in Shimonofu 2-ku, Minato and Mitoma elect two representatives each. Usually wise old men are preferred here as in the shrine, but whereas the rules for the *ujiko* representation to the shrine have been fairly fixed for a long time, they have changed considerably for the temple. Previously the representatives seem to have been chosen by the temple priest, and the criterion

Ill.13 The entrance to Isozaki shrine.

has been wealth and influence. In 1896 there was only one fisherman among the five temple representatives. Among the representatives were Sakai Manshichi, Toyomura Heishirō and Kaneuchi Shinshichi; the fisherman was Yayama Gonji, a net-owner in Nakamachi. In 1925 two of the thirteen representatives came from Honmachi, both influential fishermen and leaders of net-groups, while five came from Uramachi.

The affiliation with the temple is supposed to be based on faith, not on a particular locality. While immigrants to Shingū tend to curtail their relations with their old shrines, they hesitate to do so with their old temples, mainly because of their strong emotional ties to their family graves (cf.Table 8). Many uncommitted households, i.e. newly established households that have recently moved into Shingū, however, obviously affiliate with the local temple because of the convenience of belonging to a temple in the vicinity. This urge is not fully felt until somebody within the household passes away, and the household is faced with a funeral. While a household today can very well be without any *ujigami,* it is much more difficult not to belong to any congregation that takes care of the dead and the graves.

Some newly-established branches of old Shingū households are strictly speaking not members either, as they have not paid the initial fee. When money was collected for the reconstruction of the temple, it was decided that new members should pay at least 250,000 yen, and some found this too expensive. New households without sons regarded their households as temporary units, and did not find it necessary to become parishioners themselves, but regarded themselves as members through the membership of their respective *honke* households. Thus, of the 300,000 yen one of the fishermen donated, one third came from the branch household. Households where sons had been born, on the other hand, felt sure

Ill.14 The temple Sainen-ji. The main hall (to the right) was completely rebuilt during 1974-75.

that their households would continue to exist across the generations, and these households used the opportunity to donate this sum of money.

The parishioners are in contact with the temple several times through the year; at least at New Year, the spring and autumn equinoxes, the important festival *Bon* in August when the souls return for a visit, and during a memorial service known as *ōjūya* in November, when the recently dead are commemorated. All these ceremonies are strictly speaking aimed at the household, though some of these occasions (particularly *Bon* and *ōjūya*) may create a sense of community feeling among the participants. New Year, equinoxes, *Bon* and memorial services (*hōji*) bring relatives together, and naturally the branch households more often visit their *honke* on such occasions than the other way round, since the *honke* has no ancestors on the Buddhist altar of the branch household.

The main temple of the Jōdō Sect is Chion-in in Kyōto, *chion* meaning "knowledge of *on*". There are four kinds of *on* or feeling of gratitude: (1) *on* to the parents, (2) *on* to the society, (3) *on* to the emperor, and (4) *on* to Buddha, dharma and the priesthood. Only the first may concern us here. As already mentioned, the *on* to the parents is not terminated by death, but continues in the form of certain rituals. It is believed that these rituals will secure the ancestors a smooth entry to the Western Paradise where Amida reigns. The dead persons enter the heaven in one of nine levels, which is a mirror of the nine levels in the human existence. The fishermen occupy the lowest level of mankind, together with butchers and criminals, because their occupation involves taking life. The ancestors will nevertheless, thanks to the great mercy of Amida, climb to the top and will eventually merge with Amida himself. The descendants' observation of the rituals and the invocation of the name of Amida Buddha, the so-called *nembutsu "namu Amida Butsu"*, both help to shorten the ordeals of the ancestors, as well as help the worshipper to gain merit and thus secure his own smooth entry to Amida's Paradise. Failure to perform the rituals will, on the other hand, cause the ancestors to turn malevolent, and failure in fishing was often explained by improper treatment of the ancestors.

Another consequence of the concept *on* to one's parents has been touched upon earlier, namely the duty to continue the household and to carry it to prosperity. This has had a conservative affect as it tied the population, at least the main house, to a location and occupation. Most of the fishing households today are such main households, while the branches show a higher degree of social and physical mobility. Few of the fishermen at present feel, however, any duty to continue the household occupation, a theme I will return to in Chapter IX.

A few religious associations are complementary to the two outlined above. Within the temple precinct there is a building for the bodhisattva Kannon, the God of Mercy. A religious group Kannonkō is organized around this deity, and the group has about seventy members, all elderly ladies. Meetings are held in the main temple every month on the seventeenth (except during the winter when it is too cold) when the priest preaches. Two small groups *jizōza* are centred on the two Jizō in Uramachi, and ceremonies are arranged privately on a rotational basis by each *jizōza* every month on the twenty-third. In July, on Jizō Eve, festivals with *sumō* wrestling competitions for the children are arranged by the neighbourhood groups in Uramachi for both Jizō. The worshippers are mostly women, though the *jijōza* of the eastern Jizō consists solely of men, and the western of women. This latter *jizōza* is rather active in the devotion, and during the meetings some of them have certain revelations, though none are actually shamans. They are sometimes approached by friends, but more commonly people go to a well-established shaman or *miko* in Shimonofu or to a religious practitioner (*sensei*) in Wajiro in Fukuoka City. These persons are only consulted in cases of illness and death when it is believed that *miko* can communicate with the deceased. Most people, however, feel a little embarrassed and shameful to admit communication with *miko*, and the practice may be more widespread than the impression I got.

The "New Religions" have received adherents in Shingū, and the two most successful have been Sōkagakkai and Konkōkyō. Bernier (1975) has pointed to the fact that the two new religions Tenrikyō and Sōkagakkai in Sone represent two different alternatives to the establishment. While Tenrikyō is easily incorporated into the traditional system, the Sōkagakkai is a complete break with the same. This is equally true regarding Konkōkyō and Sōkagakkai in Shingū.

Konkōkyō was started by a poor peasant, Kawate Bunjirō (1814-1883), who was possessed by the dreaded deity Konjin. This Shintō sect, which has about 500,000 adherents (Agency of Cultural Affairs 1972:244), has its headquarters in Okayama Prefecture where the founder was born. The first converts among the Shingū fishermen must have occurred shortly after 1900 when the sect was registered. Today three of the fishing households belong to this sect, but two of them are also parishioners to the temple and all are *ujiko* to Isozaki-jinja. Among the believers are some of the most respected and influential persons in Shingū, while the supporters of Sōkagakkai are more fringe members of the community. Sōkagakkai is a complete break with the traditional norms, and it is exclusive.

Ill.15 The Umibiraki festival. The swimming season is officially opened on 1 July, and all the leaders of the township are present. The ceremony, which is believed to keep away accidents, is led by the priest of Isozaki shrine.

This means that the believers are not parishioners of the temple, but more important; most of them do not regard themselves as *ujiko.* They are willing to take the consequences of losing some contact with their fellow villagers; on the other hand they enter an organization where they are firmly integrated on all levels. For villagers to whom the local community and the *ujigami* still carry some meaning, it is extremely difficult to take this step, and the only fishing household that supports Sōkagakkai halfheartedly, has not broken the ties with the local shrine.

To summarize: through the dual organization of the household that ideally consists of three generations forming a stem family, and the local associations, the individual found his social identity. This identity found expression in the community rites mostly centred around the shrine where the community tutelary deity is enshrined. By performing the communal rites, particularly the festivals during the summer (*natsu-matsuri*), the autumn festival *Ryūgū-matsuri* (the annual festival for the fishermen) and *Tamaseri* in connection with the New Year, health and prosperity were secured for the community as well as for the individuals. The religion of the fishermen, and here I find it irrelevant to regard Shintō and Buddhism as "different religions", has direct bearing on their occupation. Only through devoted worship of the different Shintō deities and respect and worship of the ancestors, can the catches be secured, and the concepts of household and filial piety have served to conserve the occupational pattern of the households.

To the non-fishermen, however, the local community with its *ujigami* and associations carry less meaning today than in the past, and in this process the fishermen have been alienated.

The Relative Isolation of the Fishermen

We saw in Chapter III that a majority of the inhabitants have their place of work outside Shingū, and that there has been a steady change in the occupational pattern. But what are the social implications of this process?

Several persons, particularly among those employed in small companies, cultivate the old ties even today, but many have curtailed most contact with the fellow villagers. In order to understand this development we have to turn to the structure of the Japanese firm.

Very frequently the company is more than a working place; it is a "social universe" for the employees. The household-like functions of the company have been recognized by several authors (Nakane 1973, Bennett & Ishino 1963, among others). The employee is usually employed for life, i.e. until he retires at an age of fifty-five or sixty. Some companies arrange holidays for the employees and their families, run kindergartens, shops, restaurants and hospitals, and even arrange funerals for the workers, and company rites are performed regularly. In short, the employee feels a sense of security in his company, and his colleagues are his friends and drinking partners. In other words: the company can be viewed as a total system comparable to the old village. One total system, the traditional local community, has been substituted by another total system, the company.

Though the households of employees usually remain *ujiko* of Isozaki shrine, the *ujigami* means less to the employee. His livelihood is not dependent on the natural powers like wind and rain, and there is thus no need for him to be on good terms with this deity. Moreover, with the improved transport facilities, renowned shrines have become within easy reach from Shingū. The deities associated with these shrines are regarded as more powerful, and frequently the local deity is no longer approached at all on certain important *rites de passage* like the *shichigosan* (lit.7-5-3, i.e. the festival for children of seven, five and three years of age) and the *seijin no hi* (the day of attaining adulthood). The local festivals also suffer, and there are very few participants to most of them.

Not only the festivals but also the neighbourhood groups and the age classes have felt the implication of the changes in the occupational pattern. Some age classes have fared "better" than others, however. The working force consists mainly of adult men and unmarried youths; in other words the household heads and potential members of the fire brigade and the youth clubs. And it is exactly among these segments of the population where the fragmentation has been most complete. Both *Seinenkai* and *Shōjokai* have been dissolved, and the fire brigade is in danger. In a community of about one thousand inhabitants there are difficulties in recruiting fifteen members, which is the minimum size for receiving public funds. The neighbourhood groups have also been affected, though to a lesser degree, probably because of the active support this institution receives from the authorities. It is in this connection worth stressing that the social activities of the neighbourhood groups are rapidly dwindling in scope and number while the formal structure is kept intact. Nevertheless, changes can also be observed in the execution of the duties *vis-à-vis* the administration. Frequently the household heads do not participate in the meetings within the neighbourhood groups; in their place they send their wives to represent the households. Similarly, the *kumichō* frequently send their wives as deputies. This is a new trend that cannot be attributed to a higher esteem of women, but as a reflection of the irrelevance of competing for prestige on the local arena for most of the men. It has nonetheless the side-effect of integrating the housewives more firmly into the affairs of the community.

With the decline of the neighbourhood groups and particularly with the disintegration of the youth clubs, the fishermen have lost important arenas for interaction with the other villagers. The absence of the youth clubs is specially felt. The fishermen explained the present difficulties in obtaining wives partly as a result of the dissolution of these clubs. The fishermen belong to a low-status group and thus start with a handicap on the "marriage market". In the youth clubs, however, they had an opportunity to meet and make friends with the opposite sex. "They . . . sit around, giggle and flirt mildly," according to Embree (1939:169). The significance of this possibility should not be underestimated in a community where the sexes are kept as much apart as possible. So-called "love marriages" *(renai-kekkon)* occur as a consequence of the acquaintances made in these clubs. Today the fishermen are more dependent on arranged marriage (*miai-kekkon*), and in this situation their low status is severely felt (cf.Chapter IX).

The fishermen have met the isolation by strenghening their fishing cooperative. As the fishermen increasingly became economically independent of capital from outside, the majority of the villagers lost the direct interest in the fisheries. The cohesion and the characteristics of a fishing community slowly disappeared, and few people in Shingū have any idea of what it is to be a fisherman. It was felt that even inside the fishing households this alienation from the "fishing culture" was under way, and the cooperative established a section for the women *(Fujinbu)*. The objectives were twofold: firstly, to strengthen the economy of the cooperative as the wives were expected to save money in the fishing cooperative, and not in the agricultural cooperative where the members of *Fujinkai* were saving; secondly, to improve the housewives' contact with the fishing business and thus increase their understanding of their husbands' work.

Several fishing cooperatives have also established associations for the children and the youth, but Shingū cooperative is too small for this. One association called *Kyūsaikai* for young fishermen up to forty-four years exists at present. Though its main function is to rescue boats and people at sea, it is incorporated as a branch of Shingū-machi Fire Brigade.

The implication of *Fujinbu* and *Kyūsaikai* is to isolate the fishermen further, but they have undoubtedly served to strengthen the ties between the fishermen. This process is likely to continue as it seems that the neighbourhood groups and the community-wide associations are still declining in importance. Perhaps the cooperative will take over the social activities formerly found in the neighbourhood groups some time in the future, and the latter will eventually emerge as mere administrative conveniences as long as the authorities find the institution useful.

CHAPTER V

TECHNOLOGY AND ANNUAL CYCLES

Fishing in Shingū is not a static endeavour. There exist great seasonal changes, both regarding the climate and the aquatic resources. The species caught do not only change from season to season, but the size of a niche fluctuates from one year to the next. We are therefore able to say something about permanent changes only if we study the exploitation of these niches over a period of years.

In order to exploit the niche the fisherman is dependent on a technology, and the natural environments put certain constraints on this technology. An improved technology may in many ways reduce the ecological limitations, as was the case when the engine became a reality, thus increasing the operational radius of the boats and reducing the days of idleness caused by weather conditions. But the technology can never be completely independent of the natural environments.

I will in this chapter explore the different technologies used, and I will start by examining the annual cycles. By doing so I hope to be able to show the changes through the year in regard to species caught and the technologies used to catch them. By technology I will mean both the equipment used and the social organization involved in the operation of these. By technique I will mean the actual way of handling the equipment.

Though very little information is available about the fisheries in Shingū during the Tokugawa period, a tentative annual cycle will be provided, in addition to the cycles of 1930 and 1975. Some comments may be desired as to why I have chosen 1930 for examination. There are two practical reasons; the data from that year are the most complete for the whole pre-war period, and a few old fishermen can still recall the time. It should therefore be possible to obtain a fairly reliable picture of the fisheries in 1930. But more important; great structural changes were affecting the fisheries in Shingū during the first decade of the Shōwa era (from 1926).

Between 1930 and 1975 great changes occurred both on the equipment side as well as on the organizational side of the fishing operations. It is particularly with the former I will deal in this chapter. New methods developed; some have survived and are still used while others were abandoned after a few years. As long as our understanding of the organization of the fisheries can be deepened, these short-lived methods will also be incorporated in this discussion.

The Annual Cycle during the Tokugawa Period

There exists a document from 1790 mentioning the main products of Shingū-ura (Isozakigū Doc.4-2 in Kitamura 1976:20). We find that among the activities were sardine fishing and sea bream netting during the autumn, winter and spring, mackerel fishing by hold-and-line (*saba-tsuri)* during the summer and autumn, squid was caught by hook-and-line during the summer, a season when also beach nets (*jibikiami*) were used.[1]

There is no indication in the document mentioned as to what kind of technology they used to catch the sardine. But this was obviously done by big beach nets not so different from those used until recently. Several sources point in that direction.

Ill.16 The operation of large beach seines. Four net-groups are shown; to the left a group has just hauled the net, while another group is casting in the background. The group to the right is hauling the net, and the three boats in the background (right) are searching for a shoal of fish. (From Fukuoka-ken 1878:23).

In 1824 a person named Masahei borrowed 1 *kan* 500 *momme* from the *tame-gin* fund to repair a sardine net (*iwashi-ami*). In the old shrine building at Isozaki shrine there are several old paintings, two of them important in this connection. On the first from 1785, showing a big beach net and some fishing boats surfing on the waves, is written "on this beach", and the picture is obviously from Shingū. The same is true for the second painting from 1834, where many whales are caught by a big beach net. The beach is full of people and whales, and the text reads: "During fishing with sardine nets on this beach".[2] Sardine was caught by two kinds of big nets; the most common was the beach net, the other was the purse seine introduced much later.[3]

Though the sardine and sea bream fisheries continued through the winter, we can assume that the weather prohibited extensive fishing during this season. Many could, however, find employment in the *sake* breweries (where they worked until World War II) and perhaps by some of the merchants in Shingū. A tentative and very rough annual cycle can then be constructed:

Spring:	sea bream and sardine nets.
Summer:	beach netting, hook-and-line fishing for mackerel and squid.
Autumn:	sardine and sea bream nets and hook-and-line fishing for mackerel.
Winter:	employment in the *sake* breweries, and occasionally sardine and sea bream netting.

At this point we should recall the service the fishermen had to do as *kako* during the Tokugawa period. One of the main duties was to protect Nagasaki harbour, and the period for this service was from the fourth to the ninth lunar month, corresponding roughly to May-October. For many fishermen the yearly cycle was therefore more like the following:

May-October: serving as *kako*
October-December: sardine and sea bream netting
December-March: employment in the *sake* breweries
March-May: sardine and sea bream netting.

More emphasis will be put on the marketing process later, but at this stage I will only mention a few crucial points regarding the sardine. It is uncertain whether the fishermen had any influence, and if they had to which extent, over this product themselves, and how much of the profit they could get. But sardine was an important product, and taxes were partly paid from the income of this fishery. Income from sardine was also used to finance the new temple during the 1740s. The individual fishing, from which the income went more directly into the fishermen's pockets, was conducted during the spring and summer when many had to serve as *kako*. This might add to the burden of the fishermen during the period under discussion.

The above account is far from complete; probably many other technologies were used. We do know that a big mullet net (*bora-ami*) was repaired in 1827 (Akaishi Doc. in Komonjo o yomukai 1976:58), that *tatsukuri-ami (tatsukuri* being an anchovy (*katakuchi iwashi)* which was dried and used as fertilizer) were repaired in 1823, 1825 and 1830 (op.cit.:41, 50, 82), and that Shingū fishermen were allowed to use tuna net (*maguro-ami*) for one season in 1843 (Kaneuchi Doc.136 in Kitamura 1976:20-21). Information on smaller, individually operated nets is on the other hand lacking, probably since the use of these nets was less restricted by the authorities.[4]

The Annual Cycle of 1930

The data for the period 1925-1940 regarding the annual catches and their value, the equipment used and the number of fishermen operating these, are quite extensive, but something must be said about the character of these data. Most of them are official statistics prepared by the cooperative, and they were obviously used by the tax office and other governmental agencies. The reliability of these statistics is therefore questionable. The quantities and values are certainly underrated, but how much? The data from 1931 and 1932 are obviously of no value as the 1931 catches are without fail twenty per cent below those of the year before, and the 1932 catches are twenty-five per cent below 1931 again. The figures from 1929, 1930, 1939 and 1940 seem to be more reliable as I have been able to compare them with more unofficial notes for the use of the cooperative. The figures should at least give a picture of the relative importance of the fisheries.

As to the data on the number of nets and fishermen, these figures have usually been copied from one year to the next, and only occasionally brought up to date. These figures will therefore tend to be overestimates of the actual number of nets and men. Again, the data are most reliable in the years when they are brought up to date, and assuming a linear decline, I have used simple interpolation to arrive at estimates for the intermediate years (cf.Table 11 below). With these limitations in mind we should be able to proceed to the analysis of the data at hand.

The fisheries allegedly started in February or March with the sand lance fishery. One boat with a crew of four or five men was sufficient. The fish were caught by a seine (*kanagi-ami*), and the fishermen told me this was the second most important fishery, next only to the sardine fisheries during the autumn. Everybody remembers the sand lance fishery with some nostalgia.

Certainly, this fishery was once very important; in 1895 the value of the sand lance catches amounted to 30,000 yen or about ninety per cent of the total catches that year (Fukuoka-ken 1905, part III:169-71). The sand lance net that was introduced to Shingū in 1877 was allegedly among the most used nets during the Taishō era (Chikuhō suisan kumiai 1917:445, Kasuya-gun 1924:445). In the early twenties the sand lance was the most important species in Shikanoshima, and 130 boats were engaged in this fishery. At this time the sand lance fishery was, however, rapidly declining.[5] In 1930 about 4.7 tons to a value of 2,600 yen were landed in Shingū. Though the catches were small during the period under discussion, almost all the fishermen kept the nets in anticipation of better times. The importance of this fish cannot be denied as it was the first catch of the year. The winter was, as we shall see, a hard and difficult time for the fishermen, and most had to live on credit. The position of the sand lance in the minds of the fishermen has also to be seen in this light.

The sand lance fishery continued till about 1965; the reason why it was discontinued is a little uncertain. The fish became less popular among the consumers,[6] while some claimed that the quantity of the fish decreased.[7] New resources became available for exploitation during this period, and this might also have lessened their incentives to continue.

Ill.17 A sand lance seine being emptied at sea (from Fukuoka-ken 1878:4).

Ill.18 A fisherman cleaning his squid nets.

The spring was also the season to catch squid by trap-nets (*ika-mageami;* for an illustration of this net, see Ill.19 below) and basket-nets (*ika-kagoami* or *ika-subikiami* as it is officially called, see Ill.18). These latter nets were filled with branches of a plant called *shiba,* and these attract the squid that come to spawn. As the whole territory controlled by Shingū was not equally well suited for these stationary nets, this was divided into sections which were allocated for one year by drawing lots. The management units consisted of one boat and from two to four people.

The sea bream has always been an important source of income, and it has been caught by many technologies. Unlike the sand lance and the sardines, it is a stationary fish that can be exploited most of the year. During the spring it was mainly caught by long line (*nobenawa*). From May the sea bream nets (*tai-ami*) of which there were two kinds, were used. One was manoeuvred to the beach where it was landed like a beach seine; this was known as *tai-jikogiami* and I will call it sea bream beach net. The other, which was introduced later, was emptied at sea and thus called *tai-okitoriami* (sea bream offshore net).

I have unfortunately never seen the sea bream net in operation, but an attempt will be made to describe the rather complicated procedure of operation (Ill.20). A long rope tied to many cork buoys is laid in a semi-circle. Many small white wooden sticks, which are kept in a vertical position by use of stones, are tied to the rope. When the rope is shaken, the sticks glitter, and the frightened fish are caught in a seine that is set outside the rope. In the case of the beach-net variant, the whole structure is manoeuvred to the shore, and at least five boats are required; two boats to pull the rope, two boats to pull the seine and one boat for directing the operations. If the net was of the offshore type, the rope and seine were closed, and the seine emptied. At least six boats were necessary for this kind of fishing.

There were several groups (*amigumi*) operating these nets. During the period under discussion the number of these net-groups decreased from six to three, each with a crew of thirty persons. Just after World War II three groups operated the larger sea bream offshore nets for a brief period. The beach net has long been used in Shingū, at least since the Tokugawa period, and this net has been among the most important ones in all the villages in Kasuya County.

Conflicts between the groups could develop as not all the fishing grounds were equally productive. Small segments of the fishing territory were classified and grouped together in areas in such a way that each area contained both good and less attractive fishing grounds. Each group was allocated an area by drawing lots annually. When the net was taken to the shore, care had to be observed not to infringe on the other areas.

The prevalent methods during the summer were the hook-and-line fishing (*tsuri*) and different kinds of small nets. The first was used to catch sea bream, black porgy, horse mackerel, mackerel, pig fish and squid, the latter for squid, yellowtail, pig fish, flounder, small *kanpachi* known locally as *neri,* small sea bream (*kodai*), a rockfish *mebaru (Sebastes inermis)* and *konoshiro.* Most of these methods required only one boat with a small crew, though some of the nets compelled cooperation of a couple of boats. This was particularly so regarding the yellowtail nets (*yazu-ami*).

The fishermen continued to use small nets in the autumn, but the most important methods during this season were the purse seine to catch sardine and large beach nets for sardine, yellowtail and Spanish mackerel. Practically everybody participated in these endeavours, divided into large net groups.

There were two groups with about 140 fishermen operating the purse seines. The seine (*kinchakuami*) was introduced to Japan in the 1880s, and a revised type, the *aguriami,* was adopted by the Kujūkuri fishermen in 1887 or 1888 (Kujūkuri

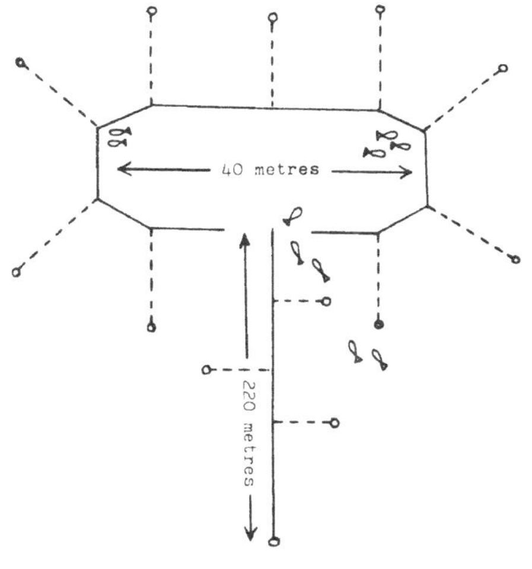

Ill.19 A small trap-net (mageami). Such nets were used to catch both squid and konoshiro, and the nets differed only slightly in size.

Research Committee 1958:27). This net was brought to Shingū from Kujūkurihama ten years later by one of the leading fishermen in Shingū at the time, Yayama Asaemon.[8]

The income from the purse seine fisheries must once have been substantial, but they were expensive. (Each net was valued to 3,500 yen in 1925.) Moreover, the catches were unstable, and the prices for sardine fell to a third from 1925 to 1932. The nets were used only occasionally since 1924 when four groups using large beach nets were established (Table 11).

These beach nets were not new to Shingū. Large beach nets were used to catch sardine as far back as in the Tokugawa period, and it may seem a step backwards that these nets again became attractive. Though the largest groups employed more than forty people, only a few boats were necessary. One small boat supervised the whole operation. Two larger boats (*amibune*) carried the net to the shoal, and tried to surround the fish as the boats separated, each throwing one end of the net into the sea. Ropes were fixed to the nets and taken to shore where most of the people were waiting to pull in the seine (cf.Ill.16). Even women with their babies tied to their backs pulled the nets. The nets could be quite large; one kind was called *senpiroami* (lit. 1000 fathom net) indicating the big size, another beach net with

Table 9 The Annual Cycle of 1930

Season	Species	Equipment	No.of boats per unit	No.of pers. per unit
Spring (Feb.– Apr.)	sand lance	sand lance seine	1	4-5
	squid	trap-nets	1	2-3
	squid	basket-nets	1	4
	wakame	collecting	1	1-3
	sea bream	long line	1	4
Summer (May– Aug.)	sea bream, black porgy, horse mackerel, squid, mackerel, pig fish	hook-and-line	1	3-4
	sea bream	sea bream nets	5	30
	yellowtail	gill nets	1-2	4-8
	small sea bream	dragnet *teguriami*	1	2
	konoshiro	trap nets	1	2
	rockfish	a seine *setsukiami*	1	3
	kisugo, konoshiro	small beach nets	1	4-7
Autumn (Sept.– Nov.)	sardine	purse seines	2	60-80
	sardine, yellowtail,	large beach seines	3-4	30-50
Winter (Dec.–Feb.)	snails, shells, sea-cucumber	collecting, digging	0	1
	- do -	fish spear	1	1-2
		young men: work in *sake* breweries old men: repairing/making nets		

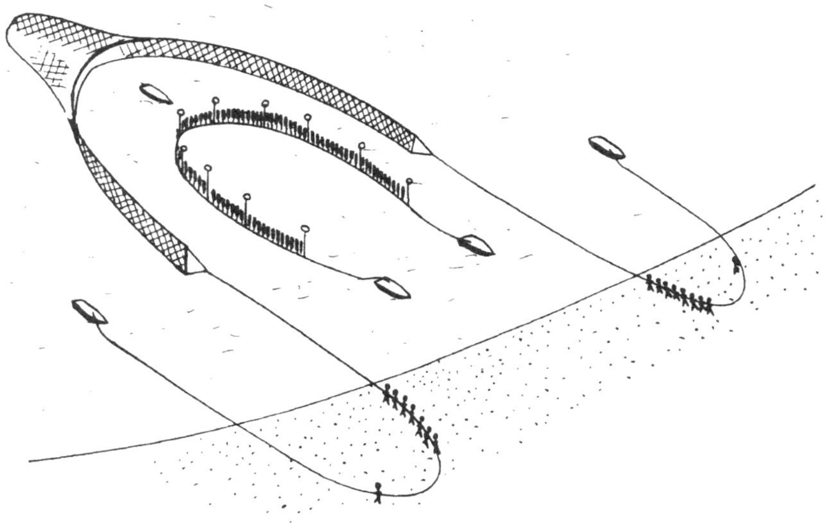

Ill.20 The operation of sea bream beach net (from Seiko 1966:70).

smaller meshes used to catch anchovy (*katakuchi iwashi*), was shorter. The net-groups operated both kinds.

The autumn was the most profitable season, but everybody was entitled to a share, so the money did not last long. The winter came and very little fishing could be done; by the New Year many of the fishermen had to borrow money. The young men were forced into employment on land, and the *sake* breweries attracted many. The older men stayed behind mending the nets and making new ones from self-made threads. In addition they collected what they could find along the beach such as shells, sea-cucumber and sea-snails. It was a spare time activity though it gave the household members valuable extra protein, but it meant little to the economy.

I have in Table 9 made a list of activities through the year. I have cited what seems to have been the usual number of boats and persons involved in a management unit. I have also included Table 10 and Table 11 to give an idea about the relative importance of the different fisheries and seasons. Table 10 shows the catches in Shingū during the eleven first months of 1929. The figures for December are lacking, but they have probably been small.[9]

Table 11 shows us the relative importance of the technologies used between 1915 and 1935. The table further shows the decline in the number of management units and people involved in the fisheries. The limitations of these data which are based on the annual reports of the cooperative, should be kept in mind when reading these tables. Though the figures may be taken as rough estimates only, they do nonetheless clearly indicate a trend.

It is clear that the number of management units and people engaged in fishing declined steadily. If we turn to the number of boats during the same period, we find the same trend (Table 12). These figures give the total number of boats, but many of the small boats were not used. We do know that in 1925 eleven of the sixty-five unpowered boats were not used, leaving fifty-five small boats for use in

Table 10 Catches in Shingū in 1929 (rounded off to the nearest yen)

Species	Jan	Febr	March	Apr	May	June	July	Aug	Sept	Oct	Nov	Total
sardines	1055	109	–	–	–	–	–	–	737	–	–	1901
sardines & others	–	–	–	–	–	–	–	–	–	1317	–	1317
squid	–	–	54	1164	635	–	–	–	–	–	–	1853
pig fish	–	–	–	46	–	1277	305	2394	665	–	–	4687
pig fish & barracuda *(Sphyraena japonica)*	–	–	–	–	–	–	–	–	–	806	–	806
sea bream	–	–	–	–	820	228	–	–	–	–	–	1048
mackerel	–	–	–	–	443	387	1031	–	–	–	–	1861
horse mackerel	–	–	–	–	–	–	–	–	162	–	–	162
yellowtail	–	–	–	–	–	–	1396	–	–	–	852	2248
other	44	47	67	420	2313	4062	3212	802	2152	–	589	13708
Total	1099	156	121	1630	4211	5954	5944	3196	3716	2123	1441	29591

the ordinary fishing. (The net-boats were only used in the operations of large nets like the purse seine and beach seines.) During the same year there were fifty-four units using the *konoshiro* net and the dragnet *teguriami*. Furthermore, there exists a document from 1930 stating that there were forty-one tax-payers (*nōzeisha)*, i.e. persons who owned boats and equipment, in the cooperative, while there were thirty-seven units operating the *teguriami* and thirty-two the *konoshiro* net. It seems that almost all the boats were equipped with these two nets in addition to hook-and-line. It does not follow, however, that all these technologies were used

Table 11 Number of management units and their catches between 1915 and 1935 (catches in kilograms)

Equipment		1915	1920	1925	1930	1935
sea bream net	units	6	6	5	4	3
(tai-jikogiami)	people	180	165	150	120	90
	catch	7875	3000	9375	6750	4690
sand lance net	units	53	40	30	30	30
(kanagi-ami)	people	241	160	120	120	120
	catch	14625	3750	–	4680	–
yellowtail net	units	50	32	20	20	20
(yazu-ami, buri-ami)	people	200	109	60	60	60
	catch	–	1875	11250	17810	3040
flying fish nets	units	27	10	9	8	6
(tobiuo-sashiami)	people	66	30	27	24	18
	catch	4850	2625	4500	1250	–
konoshiro net	units	69	60	54	40	32
(konoshiro-ami)	people	138	120	108	80	64
	catch	2250	2250	9375	6000	195
squid trap-net	units	29	23	15	15	15
(ika-mageami)	people	87	69	33	41	45
	catch	10500	–	–	3750	9490
squid basket-net	units	35	39	45	30	25
(ika-subikiami)	people	140	147	180	120	100
	catch	6750	11250	11250	5250	5810
a seine	units	23	18	25	18	18
(setsukiami)	people	92	59	100	54	54
	catch	7500	3375	7500	5060	4875
a small dragnet	units	73	60	54	40	37
(teguriami)	people	146	130	108	80	74
	catch	13125	3750	–	9000	560
long line	units	3	10	20	20	20
(nobenawa)	people	12	40	100	80	80
	catch	6750	2625	6000	20625	560

Table 11 (continued from previous page)

hook-and-line *(ippontsuri)*	units	80	60	60	40	35
	people	240	220	180	120	105
	catch	9000	57000	4875	36000	8820
gill net *(tateami)*	units	7	5	2	2	2
	people	28	20	10	8	8
	catch	525	1500	750	1875	250
sardine purse seine *(iwashi-aguriami)*	units	2	2	2	2	2
	people	142	160	154	140	120
	catch	37500	7500	7500	37200	77625
large beach seine *(ōjibikiami)*	units	–	–	4	4	3
	people	–	–	180	133	120
	catch	–	–	26250	7950	75375
mullet net *(bora-ami)*	units	–	–	–	–	2
	people	–	–	–	–	120
	catch	–	–	–	–	18750
tatsukuri net *(tatsukuri-ami)*	units	11	–	–	–	–
	people	176	–	–	–	–
	catch	12750	–	–	–	–
tatsukuri purse seine *(setatsukuri-ami)*	units	2	20	–	–	–
	people	20	100	–	–	–
	catch	15750	7000	–	–	–
hachida net	units	2	3	2	2	2
	people	120	180	116	94	80
	catch	–	–	–	–	–
Total catches		149750	107500	98625	163200	210040

each year, but on this aspect we have unfortunately no information from this period.[10]

If we return for a moment to Table 11, we will notice two nets not previously mentioned, namely the mullet net and the *hachidaami* (or *harikuriami*). Neither of these nets were important economically during the period, but they are important in order to understand the scope of cooperative fishing and the position of net-groups in the consciousness of the fishermen. In 1915 two more kinds of nets existed though no catches were reported. These nets were medium-size beach nets (16 units engaging 160 persons) and large beach nets (7 units engaging 210 persons) both used to catch sardine. A dichotomy is clearly seen in the size of the management units; small units of a few persons co-existing with large structures. This pattern could also be observed in 1975, though not so marked.

The Annual Cycle of 1975

The *gochiami* fisheries are by far the most important in Shingū today. This net resembles the Danish seine and consists of a bag with rather long wings. The seine

Table 12 The number of Boats in Shingū 1920-1935

	1920	1925	1930	1935
Small boats without engine	73	65	42	37
Small boats with engine	0	1	5	8
Large net-boats *(amibune)*	4	7	6	4
Total	77	73	53	49

is tied to two long ropes, and one rope-end is fixed to a buoy before the ropes and seine is cast in a circle. When the boat has returned to the starting point, the buoy is picked up and the boat proceeds until the seine is closed. The engine is then used to haul in the seine, and the boat is steered by an underwater sail, the size of which depends on the strength of the tide. When the net is brought to the boat-side, it is lifted on board by hand. A winch to do this job is developed, but is at the moment regarded as too expensive.

Though it is possible for one man to operate the net alone if the sea is calm, two men are regarded as the minimum for this fishing in Shingū and three as the ideal, though in the neighbouring community, Nata, they use both four and five persons on the boat. The whole operation of casting and hauling the net can be done in about fifteen minutes, and a crew can in one day cast the net more than thirty times under favourable conditions; which depends on the fishing grounds, the weather and the size and skill of the crew.

There are twelve boats in Shingū doing this fishing, and the main catches are *kisugo* and sea bream fry. The latter is sold in two conditions; alive and dead. The dead, which is locally called *mame,* has a certain value though usually not as high as the value of *kisugo.* As both the fishing grounds and the techniques used to catch *kisugo* and the fry are a little different, there arises sometimes a conflict whether to go for one or the other. The living fry of sea bream is, however, a highly valued product and over-shadows everything else. This fishery is called *taichigyo* fishery.

The *gochiami* fisheries start on May 1 and can continue through January, but nobody continues that long. In November the catches usually fall sharply and the weather becomes too bad, so the fishermen switch to something else. The season for the *taichigyo* fishery is also decided by the authorities, and the season depends on the size of the fry. It lasts for only a few weeks during the height of the summer, when the fry is about two inches long. This fishery is a cooperation between all the boats operating *gochiami,* and as it is the most important cooperative endeavour, its organization will be discussed at length in Chapter VII.

Only a few fishermen do not participate on the *gochiami* boats. Common to all of them is that they operate small nets mostly alone. Two young men fished for *kisugo* by drifting trammel nets (*kisugo-nagashiami*) for a couple of years. The season for this is the same as for *gochiami.* (These units are *B2* and *B4* in Figure 1 and Table 17.) One of them participated, however, in the *taichigyo* fishery during the summer 1975, while the other in 1976 joined a crew on a new *gochiami* boat. One man (*B5*) in his forties operated different kinds of small nets to catch ivory

Ill.21 A small beach net. The fisherman is hauling a beach net cast on order by tourists.

shell, squid and the like. He participated in his brother's boat during the *taichigyo* season and bought a new boat with him and *B4* above. An older man (*B1*) operates many different nets throughout the year. He is mentally disturbed, divorced and lives alone. He is in many ways a marginal member of the cooperative. One retired fisherman whose son is working on a *gochiami* boat conducts small-scale fishing with his wife and brother-in-law (*C2*), using gill nets to catch flying fish in the spring and small beach nets during the summer. This latter is primarily done on orders from tourists who, during the summer, come in groups to Shingū to have some fun. They hire the fishermen and participate in pulling the net. The catch is theirs and is normally consumed at the following party which takes place at the beach. (This has been a new niche since the war, and many fishermen from Ainoshima take anglers on board their motor boats; a practice disliked in Shingū.) Another man (*C1*) does the same kind of business with his wife, and he also does some other small-scale fishing. He is the only one who still carries on the squid trap-net fishery once so popular. The brother of the cooperative's manager (*B3*) operates small gill nets for flounder and crabs during the summer and autumn; he collects shell and sea-cucumber during the winter and *wakame* during the spring and early summer. The last person to be mentioned here is a former fisherman who established his own plumbing business. Since 1976 he has given this business to his son, and he has returned to the fisheries where he uses several kinds of small nets.

There are two molluscs which are particularly sought during the winter. The most important is the ivory shell, which is caught in small traps; mackerel is used as a bait. The trap is formed like a basket, formerly made of bamboo but now made of an iron frame and net. The ivory shell climbs along the edge of the trap to reach the bait, but falls inside the net, a simple but extremely effective device. One boat may have as many as two or three hundred of these, and they are emptied every second or third day. One person can do the job, but two can cover more. Seven

Ill.22 November is the season to change to winter activities, and here two brothers are preparing the octopus-pots.

boats were engaged in this activity during the winter of 1975/76. Four other boats caught octopus with terra cotta pots (*tako-tsubo*) that resemble flower vases. These are tied to long ropes and lowered to the bottom of the sea. The octopus is attracted by the dark entrance and remains inside the pot until the fisherman has lifted the rope on board. An improved version was introduced in 1974 by one of the fishermen. These are made of concrete and the entrance is closed as soon as the octopus has entered. As for the shell-trap (*bai-kago*), one man can operate the octopus-pot, but two can do the job more efficiently.

Eight boats were operating small nets during the winter, three of these were fishing grey mullet by mullet net, the other were mostly fishing flounder, crabs, horse mackerel and barracuda (*kamasu*). One man can operate these nets, though two are an advantage when fishing the grey mullet. Three persons were collecting snails, shells and sea-cucumber along the shore; they were using small rowing boats and spear.

Several boats use more than one technology during the winter. This season has always been difficult for the fishermen, and no technology has been found that can keep them active and give them a reasonable income. The profits during the winter are much less than during the rest of the year. In the fiscal year 1974 only 1.06 per cent of the total catch was taken during January and February. The fishing co-operative has tried several measures to increase the activities during the winter. An attempt to grow *nori* was a failure, while the *wakame* experiment was a modest success for a few years. Subsidies were given to those who wanted to try to cultivate *wakame.* Only three crews were interested, partly because it required a big capital investment and the returns were uncertain, and partly because the harvest begins in March and continues through May. In other words, the *wakame* does not solve the problem of idleness during January and February, and the harvest has to be stopped in May as the profitable *gochiami* season starts. The income from the *wakame* cultivation was not large, and the three crew did not re-invest in this undertaking. By 1977 all cultivation had ceased.

The fishermen nonetheless accept readily innovations which are obviously beneficial. The history of the ivory shell illustrates this point in full.[11]

Many years ago a fisherman *A* was approached by a trader who asked him to catch ivory shell which he guaranteed to buy. *A* made one catch, and he used a trap made of bamboo. Unaware of how to treat the shells, he put them in the refrigerator. The shells froze to death, and the trader did not return. Many years later another trader approached *B* with the same offer. *B* made a catch with the same kind of equipment as *A* had used. This time the trader returned, and the price fetched was surprisingly high. This encouraged *A* to make another trial, and he went to some farmers to order more traps made of bamboo. But nobody was willing to make these any more, and he travelled to many villages in the vicinity. Finally he came to a fishing village in Hakata Bay, and there he saw some similar traps made of wire and net. He went straight home to make some, and he received a spectacular catch. The following week most of the fishermen were sitting the whole nights through making such traps. The beating of iron (which goes *kachin-kachin-kachin* according to the local onomatopoeticon) could allegedly be heard from all the houses. The next year there was hardly any ivory shell left. Today the situation has stabilized; the catches are smaller, which implies a lower profit, and many

Ill.23 The planting of *wakame*. In November the fishermen received the seeds which are grown on thin ropes shown here. These ropes are later twined to thicker ones and lowered into the sea.

have returned to technologies like the octopus-pot and nets. This technology is nevertheless still the most important during the winter.

Though most of the boats managed to continue operations during the winter months, they did so with reduced crew. Two *gochiami* boats were inactive during the winter 1975/76 (*A5* and *A9* in Fig.1), and between eight and ten fishermen took employment in other occupations, that is in construction works, which is the only available seasonal job for unskilled labour. On the other hand three persons usually employed on the *gochiami* boats operated their own small crafts during the winter and spring *(D1, D2* and *D3). (D1* has, since 1977, been in operation the whole year as the crew has ceased to participate in the *gochiami* fisheries.)

In March the season for squid starts, and most of the boats take part in this. The method is described in the previous section, and also here one man is now sufficient, strictly speaking. This activity continues until the *gochiami* season starts in May. During March and April many of the people also collect natural grown *wakame.*

It may seem that cooperative efforts had little importance in 1975. Only the short *taichigyo* season calls for cooperation, and even here there is no cooperation on the sea, only in the marketing and feeding of the fry. Nonetheless there were groups in existence "under the surface", and these could be activated. Yellowtail has, as we have seen in the last section, played a certain role in Shingū, and it still does. This valuable fish is often caught by two gill nets which are laid around a shoal of fish. Two boats are needed in order to cast the nets, but more boats are sometimes required to carry the catch to the harbour. There were in 1975 four groups owning such nets, three of these were remnants of former big net-groups. When I revisited Shingū in 1978 two of these three were dissolved, and cooperation between boats in this fishery will in the future be made *ad hoc.*

One net-group, comprising all but a few fishermen, was dissolved in June 1975 after four years of operations. This group owned four large stationary trap-nets

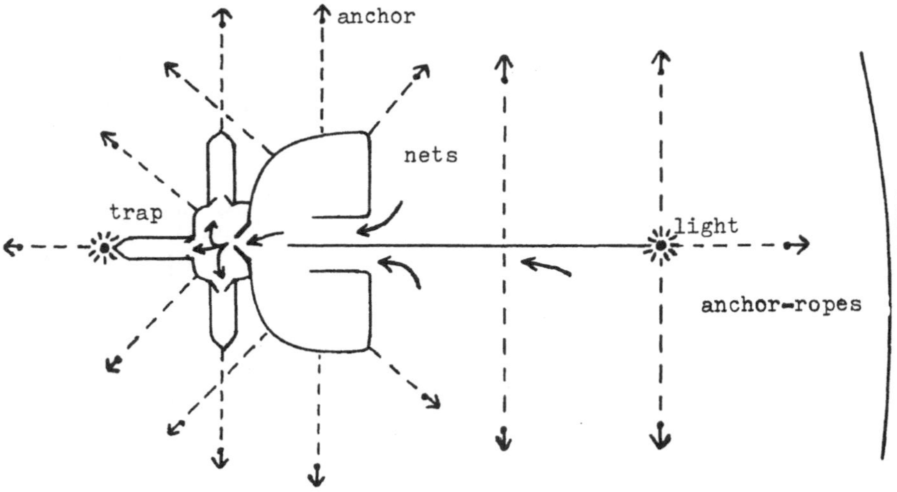

III.24 The large stationary trap-nets (*teichiami*).

Table 13 The Annual Cycle of 1975

Season	Species	Equipment	No. of boats per unit	No.of pers. per unit
Spring (Mar.- Apr.)	squid	basket-nets	1	1-3
	wakame	collecting	1	1-2
	wakame	cultivation	1	2
	flounder, squid	gill nets	1	1-3
Summer (May- Aug.)	*kisugo, mame*	*gochiami*	1	2-3
	living sea bream fry	*gochiami*	1	3
	kisugo	trammel nets	1	1
	squid, *hirazu*	large trap-nets	2	5
	flounder, crabs	gill nets	1	1-2
	ivory shell	traps	1	1-2
	miscellaneous	small beach nets	1	2-3
	egonori	collecting	0	1
Autumn (Sept.- Oct.)	*kisugo, mame*	*gochiami*	1	1-3
	kisugo	trammel nets	1	1
	horse mackerel, barracuda, *hirame*	gill nets	1	1
	yellowtail	two gill nets	2-4	4-8
Winter (Nov.- Feb.)	ivory shell	traps	1	1-3
	octopus	octopus-pots	1	1-2
	sea-cucumber, shell	fish-spear	1	1
	mullet, flounder	small nets	1	1-3
		Construction works		

(teichiami). The nets were emptied three nights a week, and the owners took turns so that everybody had to participate once every week. The group was divided into three sub-groups which again consisted of two teams. Each team emptied two nets during the night. Four boats were required; two net-boats from which the nets were emptied and two powered boats to pull these. Every second month they had to take up the nets for cleaning and dyeing, and they had to hire professionals to place the nets afterwards. These nets, which operated from April to October, have been sold.

We should by now be able to summarize the annual cycle as it was in 1975 (Table 13).

Table 14 shows the relative importance of the different species caught in Shingū in 1976 according to month. The relative importance of the different technologies in terms of weight is expressed in Table 15. The three last years have been included. The catches were markedly more stable than in the 1930s, partly due to improved technology, partly since the most important species at present are stationary fish, not capricious pelagic species.

Table 14 Catches in Shingū in 1976 (in kilograms)

Species	Jan	Febr	March	Apr	May	June	July	Aug	Sept	Oct	Nov	Dec	Total
octopus	396	86	–	194	–	–	–	18	493	–	2388	2987	6562
ivory shell	870	430	40	255	240	625	315	–	45	–	–	1980	4800
grey mullet	142	110	56	–	–	–	–	–	–	70	98	–	476
turban shell *sazae; (Turbo cornutus)*	30	–	15	–	–	–	–	–	–	–	–	40	85
abalone *awabi; (Haliotis discus)*	20	40	15	40	–	–	–	–	–	–	–	–	115
flat-fish *(karei* and *hirame)*	–	624	3946	1492	–	20	–	–	355	1885	27	77	8426
squid	–	–	2669	5030	910	1824	–	–	–	–	–	–	10433
kisugo	–	–	–	–	7359	2327	3681	2640	4826	10274	2653	128	33888
dead sea bream fry	–	–	–	–	1548	587	3277	3529	813	476	147	21	10398
living sea bream fry	–	–	–	–	2525	6000	5730	–	–	–	–	–	14255
yellowtail	–	–	–	–	–	150	2845	–	–	–	–	–	2995
barracuda	–	–	–	–	–	–	–	15	–	960	–	–	975
horse mackerel	–	–	–	–	–	–	–	–	529	10	–	–	539
eel *anago; (Astroconger myriaster)*	–	–	–	–	–	–	–	–	36	–	231	297	564
other	–	–	–	–	1377	2426	2239	1010	542	350	101	240	8285
Total	1458	1290	6741	7011	13959	13959	18087	7212	7639	14025	5645	5770	102796

Table 15 Catches in Shingū by technology (in kilograms)
Fiscal Years (April-March)

Equipment	1975	1976	1977
gochiami	80,885	77,741	66,146
gill nets	6,368	12,463	15,696
squid basket-net	6,094	7,867	7,320
ivory shell trap	7,177	7,457	3,822
octopus-pot	3,255	2,091	2,178
squid trap-net	136	123	181
kisugo trammel net	1,306	346	38
large trap-net	3,704	–	–
mullet net	1,355	740	–
yellowtail net	296	1,190	3,145
sand lance seine	1,379	–	–
small beach seine	18	239	271
hook-and-line	–	70	614
octopus basket-net	–	–	6,032
fish-spear	140	165	205
wakame cultivation	612	–	–
other	–	40	66
Total	112,725	110,532	105,714

Summary

Some features stand out sharply if we examine and compare the annual cycles for the Tokugawa period, 1930 and 1975. First, the dynamics of the fishing activities through the year: the management units are changing from a single man on a boat to large units of thirty or more men on several boats. This feature was particularly strong in the past when big units conducted a large part of the fishing. It is less apparent in 1975. The large group operating the trap-nets was dissolved, and the yellowtail net-groups were smaller and only rarely active. But this impression does not represent the full picture; we have to examine the procedure regarding the *taichigyo* fishery carefully before we can reach any conclusion as to the importance of cooperation today. There is nothing in the technology of this fishery that requires the cooperation that actually exists between the twelve boats. They operate independently on the fishing grounds but pool their catches before marketing. The profit is divided among all the boats regardless of how much each contributed to the common catch. The impact of this fishery is therefore the same as other cooperative activities as we found them in the past; the most important income in the community is distributed almost equally among the participants, thus levelling an otherwise uneven distribution of income. The income of the household was, and is, directly dependent on how many members it contributed to these endeavours.

Secondly, the feature of great changes over time is apparent. The niches exploited and the technologies used have changed considerably during the last forty years. Furthermore, the fishing population has become more homogeneous today as the vast majority is engaged on the *gochiami* boats. The others are in many ways marginal members. In 1930 there was, on the other hand, a clear distinction between those who mainly used hook-and-line and those who used mainly nets, and I will later correlate this distinction with membership of the net-groups we could find at that time, and with the viability of the fishing household. Today nobody is using hook-and-line any more except occasionally, more or less for fun. Did then many fishermen change to nets during this period? A few did; it is illuminating that those using hook-and-line were the last to turn to *gochiami*. But most of them have left the fishing business and taken other jobs. (For an analysis of this process, see Chapter IX.)

Two technologies used for a few years between 1930 and 1975 should be mentioned as they represent technologies linking those in 1930 and 1975. The first was a kind of seine called *nuikiriami*. A boat with a torch-light attracted the fish whereupon the boat gave a signal to other boats to cast the net. The net had big meshes except for a purse at the end which was fine-meshed and formed like a trap. When the net was in position, two boats took the torches and rowed towards the trap, and the fish followed. The two net-groups operating these nets used ten boats and engaged eighty people. The season was from early summer to the middle of September, and for a successful warp there must be no moon nor wind.

The second technology I want to mention here is the *gochiami* operated by two boats *(nisō gochiami* as opposed to *issō gochiami)*. It worked as the *gochiami* already described, but the buoy was not necessary; the two boats pulled the net for a short time like a trawl. As the seine was hauled by hand, about five men were needed on each boat, and the income for each was not large. Many tried this technology in Shingū, but failed and soon gave up. Four or five pairs of boats managed to continue for some years before they slowly changed to *issō gochiami*. The hauls consisted mostly of sea bream as they had still not developed the technique to catch *kisugo*.

The main technologies which have been and are being used in Shingū have now been described. The management units can, on the basis of the number of boats participating in a fishing operation, be divided into two groups: (1) the single boat which operates independently on the fishing grounds often in conflict with other similar units, and (2) cooperation with other boats, either in what I will call net-groups or in marketing the catch.

Both kinds of management units have existed side by side as long as we have records from Shingū. Though the structures of these two kinds of units can show parallel development in terms of egalitarization, they need nonetheless separate treatment. The former, which will be analysed in the next chapter, has mostly been a matter of free initiative, i.e. the fisherman himself. The latter, on the other hand, required not only capital outlays which the fishermen did not have, but also a licence and a more developed market organization. It is in the structures of the net-groups that the most radical changes have occurred, and this will be the subject of Chapter VII.

CHAPTER VI

THE FISHING BOAT AS A MANAGEMENT UNIT

The Concepts of Status and Resource

In the preceding chapter I tried to stress the great flexibility in the size of management units in Shingū by focusing on the annual cycles. This flexibility was particularly in evidence in 1930 and to a lesser degree in 1975. Nonetheless, the same principles for the flexibility also existed in 1975. Even by looking at one single boat, we can find these changes through the year. It is the same boat that was operated by one man in January 1975, by two men in May and by three in June. The changing size of the crew can be explained partly by ecological and partly by cultural factors, but at this stage I will limit the analysis to the exploration of the different statuses we can find in the management units.

It should be emphasized that by status I mean a set of rights and obligations of an actor (Linton 1936). All the persons on the boat can be described as fishermen, and they have a status as such. By being a fisherman, he has the right to fish inside the territories of his cooperative, and he has the obligation to be a member of this cooperative. As a fisherman he has the right and duty to sell all his products through the market mechanism decided by the cooperative, and during the Tokugawa period he had to serve as *kako*. The status as a fisherman does then have a great significance in his daily life. However, the status "fisherman" is hopelessly inadequate when I want to discuss the interactions that take place on the boat where everybody has this status. It is therefore necessary to make a further distinction; a person had several statuses, and I will follow Goffman (1971) and Barth (1966) in that an actor's relevant status depends on the situation which has to be defined in agreement with the actors in the interaction. Gerrard has analysed the fishermen in a North Norwegian fjord in terms of three statuses: the skipper, the experienced fisherman and the beginner (Gerrard 1975). This has proved a very useful approach, and many of the rights and obligations that can be isolated are common to that fjord community and Shingū. I intend, however, to focus on the changing size and composition of the management units found in Shingū, and the three statuses mentioned above are only partly relevant for the understanding of these changes. It is irrelevant to the fishermen in Shingū whether they are experienced or not when the crew is reduced; there are other factors at work. I will therefore not be able to explain the flexibility found by using the set of statuses outlined by Gerrard. Nor will I be able to analyse the leadership patterns found on the boats.

It is possible to divide the total personnel on a boat into three categories: the skipper (who is also an owner of the equipment), the "core crew" and the "extra crew". The skipper and the core crew form the personnel on the boat most of the year. The extra crew enters the boat only during certain seasons when additional crew members are wanted. The skipper alone on the boat is a rare sight except during the slack winter season when the core crew often has to find other employment, either in construction work or on their own boats. (It should also be clarified that there are boats where only the skipper works the whole year through, but they do not concern us here, as we are discussing the interactions between statuses on a boat.) My set of statuses will be associated with these three categories of personnel; the status as skipper, as member of the core crew and as member of the extra crew.

I will in the following analysis also use the concept "resource". For each status there are resources that can be activated during interaction, and which can be important for the role performance of the actors. By a resource I will mean any activum that can be utilized in a transaction; material objects, knowledge, working power, social relations etc. It becomes one of the main tasks in this and the following chapter to isolate the resources relevant to a status, and to see how these resources are distributed and used in interactions with others.

I have so far introduced three statuses which can be found on a fishing vessel. Other statuses can also be isolated, for instance the status as owner which carries many specific rights and obligations. The skipper is, however, in most cases an owner, but he might share the ownership with other persons either on land or with members of the core crew. I will treat this status as a resource for interaction; a resource that might strengthen or weaken the skipper's basis for action. Other statuses can be treated in the same way. Those most relevant for the role performance are the status of man or woman, experienced or inexperienced fisherman, old or young.

I will in this chapter try to show that where the personnel on a boat possesses different resources, a hierarchical crew will develop, and the skipper will seek to tie crew-members to himself in patron-client like relations. If, on the other hand, the crew-members possess more or less the same resources as the skipper, the latter has to invest more in his relations with his crewmen in order to keep them, and the relationship tends to be more fragile and purely economic in character.

It became clear from the preceding chapter that there exist structures where more than one boat is involved. The statuses found in these structures are often different from those found on a boat. Though we in many cases are still able to talk about a skipper, core crew and extra crew, the resources, duties and rights associated with these statuses are different. I will therefore treat them separately, and this I will do in the next chapter.

The Status as Skipper

A person needs some resources in order to be a skipper, and the single most important one is the ownership of a fishing vessel. He does not need to be the sole owner, but there are few cases in Shingū where the skipper does not own any part of the boat. There were in 1975 only two cases, both involving sons who are skippers on their fathers' boats, and both had by 1978 succeeded as owners. He then needs access to capital either as savings or as loans; in the latter case he has to be regarded as credible by the financier. The other absolutely necessary resource is the membership in the fishing cooperative, either as personal member or as a member through his household. Only those who belong to households of members are entitled to be skippers. The membership is heritable, but can also be obtained by purchase of shares (Chapter VIII).

The skipper furthermore has to possess some knowledge of the technique in the fishing operations, on the natural conditions on the fishing grounds, on the different species of fish and on the fishing regulations of the area. He also needs licences from the fishing authorities in Fukuoka. These are of two kinds: (1) A certificate that proves he is able to sail the boat in question. To obtain this licence he has to follow a course and prepare for a test, which so far has been easy for the fishermen to pass. The whole procedure has been done *pro forma* as the intention of the regulation is to restrict the owners of leisure boats. In fact, almost all the active fishermen in Shingū have taken this licence, at least the lowest grade which entitles them to steer the boat of this size. (2) He needs a licence to conduct a specific kind of fishing, and the authorities are more restrictive on this to avoid overfishing. Many kinds of fishing, including *gochiami,* need to be licensed.

Though the access to these resources might have been limited in periods, they do not put serious restrictions on the fishermen today. The crucial factors at the moment are: to secure labour and to possess self-confidence. As to the first, if no labour can be found except one's own, one can still continue to fish alone. A few persons do this, and more may have to do so in the near future. The fishing population is decreasing steadily, and several of the boats using *gochiami* have no recruits in sight. As to the second crucial resource; all the fishermen stressed the importance of self-confidence. If this is lacking, the role performance of the skipper will be poor, he will be inclined to ask his own crew for advice jeopardizing his own position. He will also be conservative and hesitant in adopting new technologies. The only *gochiami* boat which has stopped operations so far, stopped allegedly for this reason. The boat that was operated by one of the most influential fishermen, was not among the poor catchers, and his son participated. Nonetheless, on the death of the father, his son failed to continue.

These are the resources the skipper needs in order to exploit an aquatic niche, and these resources are utilized to meet his obligations as skipper. His most important duty is to see that the whole fishing operation goes smoothly, and that the participants can get a reasonable return for their troubles. More specifically, he has to decide whether to fish or not. The weather conditions have to be considered; he has to be concerned with the safety both of himself and the crew. But even when the weather is excellent, the skipper might decide to remain at home. Not infrequently I heard that the fishermen stayed at home because the prices at the markets were low, the fishing had been bad lately, or they were simply tired. The decision to fish or not is not only the duty of the skipper, it is also his privilege. It is he who decides whether the crew is exhausted or not, though everybody may have a word in the decision. He is also greatly influenced by the other skippers (cf. below).

The skipper has to check that the boat and equipment are in good condition; every morning he has to be the first at the harbour to get the boat ready. Fuel has to be tanked and ice and baskets have to be loaded on board. When the boat returns in the afternoon, he washes it and secures the boat at a buoy in the harbour. During typhoons he has to decide whether to land the boat. If anything goes wrong, he has to buy spare-parts, repair the damage if possible, or to call a specialist. He has to plan the repairs and other maintenances so as to minimize the dead days; he is obliged to keep the boat in business as often as possible. He likewise has to keep the nets in good repair, to mend the old or to buy new ones. The *gochiami* nets are often broken due to rocks and wastes on the bottom, and during dead days he is usually found mending his nets on the beach or in his own house.

Ill.25 Boats at the harbour in the morning. Usually all the boats leave the harbour within a few minutes.

The skipper has the obligation towards the rest of his crew to secure enough labour for the operations and to hire additional members when required. But as the size of the crews has been decreasing, this obligation becomes less relevant. The problem now is rather to get a second person on board. If he cannot gather sufficient crew, they may change the technology though there is a certain degree of flexibility in this respect, partly due to mechanization.

On the fishing grounds he has to coordinate the activities. He is obliged to bring the boat to a good catch, his success is valued both in terms of income from the catch, but more important in relation to the result of the other boats. In order to improve the performance of the boat, he has to utilize all his knowledge about the fishing grounds, the tides, which species are present and so forth. Not all the fishing grounds are equally good, and he has to guess which will yield a good catch on that particular day. If his judgment is wrong, he has to realize this quickly and go to an alternative spot. All the skippers have their special spots which they guard with some secrecy. This does not mean that they have exclusive rights in these areas, but he knows the spots and believes they will bring him a good catch. A good skipper knows more good spots and he is quicker to realize a mistake than are the poor skippers.

Poor catches are often explained by the skippers concerned as being consequences of improper religious behaviour, i.e. that they have offended a deity or an ancestor in one way or the other (Otō 1963:114). As one skipper said: "Whenever I have poor luck or my nets are being destroyed, I wonder if I forgot to offer food and tea to my ancestors in the morning." And many households would refrain from eating chicken and egg in order not to offend the deity Susanoo-no-mikoto (cf. note 9 Chapter IV). While this is obviously a device to put the blame for poor catches on supernatural beings and thus protect one's self-confidence as a skipper, the fishermen themselves have a clear idea as to who are the better skippers, and this judgment is done solely on the criterion of relative catch.

The skipper's knowledge should also extend to the market situation – only through this knowledge can he judge properly which species they should catch. The prices on the market vary; one day it may be profitable to catch *kisugo* and the next day it will be better to catch *mame* (sea bream fry sold in dead condition). There are two factors determining the value of the catch: the quantity and the quality of the fish, and both have to be considered to optimize the profit.[1]

The skipper is also held responsible for any misconduct of his boat. The fishing grounds are covered by a wide range of regulations, and it is his duty to know these. If the rules are broken, he is held solely responsible. These regulations do often interfere with profitable operations, and he has to consider the risks involved in breaking these in relation to the profit gained. This is a problem he is faced with almost daily.

Though very few orders are given (everybody knows what to do), he has to keep an eye open, and he has to take the lead in critical phases of the operation such as casting and hauling of the net.

The duties of the skipper are many, and so are his worries. Some expressed the attitude that it would be better to be employed. The employed have the same share of the catch and no worries, they said. But still – the skippers would not change.

He has some rights making it attractive to be a skipper. First of all, he has the privilege of autonomy, he can decide when and where to go. He takes the decision what to catch and thus also the technology to be used. In the case of *gochiami* he has the right to decide whether to fish *mame* or *kisugo.* This is often a difficult question where the interests of the skipper and his crew might differ (cf. below). The same holds true when he has to decide when the winter activities start. This often results in the layoff even of the core crew. The skipper is thus the last person on the boat that has to turn to construction work which is regarded by most fishermen as humiliating for two reasons. First, the fishermen regard themselves in a way superior to ordinary construction workers; the work does not carry the same prestige as fishing. Secondly, they are allegedly harassed by the regular employed workers who, in this kind of work, are their superiors.

As the owner of the boat and equipment the skipper also has more control of the finances than the crew members. He can take some percentages away from the gross income to pay for the boat, equipment, running expenses etc. The key to the distribution of the profit may vary to some degree, but usually the skipper does not receive a bigger share than the others. Two examples will suffice to illustrate the dispersion of the catch.

Example 1: Ten per cent of the gross income is usually spent to pay for the boat either as interest and instalments on loans or as savings for a new boat. Another ten per cent is taken to cover running expenses as nets, spare-parts, oil etc. Marketing is another expense; the fishing cooperative takes six per cent and the market in Hakata another five. The fishermen also have to pay for boxes, ice, storage and transport; expenses that are more dependent on the quantity than the quality of the catch.[2] Altogether about forty per cent of the gross income is spent on these expenses, and the rest is usually divided equally among all the participants on the boat.

Example 2: One skipper had a somewhat peculiar key which he inherited from his father. It may reflect a more archaic tradition, particularly since it includes "*sake* money" (tips) to the crew. If he employs people outside the family, he will take ten per cent for the boat, another ten per cent for equipment, and fifteen for marketing. Thirty per cent of the rest is kept for running expenses. The net income is then divided equally. Three per cent of one man's share is taken from the running expenses and spent on *sake* for the crew members. It will be clear that less is distributed to the crew-members in this case than in example 1.

Though everybody receives equal shares of the net income, it will be clear that the owners of the boats benefit in economic terms. Once they have a boat they are given good opportunities to re-invest in new equipment (cf. Chapter IX). The ten per cent for running expenses is furthermore not always spent in full. Previously when the income was low, the owners sometimes had to add extra funds to cover these expenses, but the good income during the last years has changed this. A few owners divide this surplus at the end of the year between co-owners as a "bonus". We have therefore tendencies to stratifications between the owners and the other crew members during good years (cf. Epstein 1967:250, Lofgren 1972:98), but this stratification is slight.

The skipper participates in the fishing cooperative as a member, but he also participates there as a skipper. There are frequent meetings between the skippers

Ill.26 A married couple sorting the day's catch of *kisugo*.

where the others are excluded. As will be shown in the next chapter, the skipper also takes a more active role in the cooperative efforts like the *taichigyo* fishing than do the other fishermen.

The Status as Member of the Core Crew

If we compare the skipper with the members of the core crew on one of the fishing vessels in Shingū today, we might not be able to distinguish them. Actually, they may possess more or less the same resources. Thus, many crew-members own shares in the equipment on an equal basis with the skipper. It will therefore be useful in the following analysis to isolate the minimum of resources necessary to hold the status as a member of the core crew. The resources a member might have in addition are irrelevant to the status as a crew member, but they do of course become relevant when we consider how he performs the role as a member of the core crew.

The most important resource the member of the core crew possesses is his own working power, and his foremost obligation is of course to utilize this power. He does not get any written contract, but his responsibilities are clear. The relations between the core crew and the skipper are stable and can continue through many years. The member is tied to the skipper with a multitude of mutual obligations. He has to participate on the boat when the skipper decides to fish. If he is absent for a long time without good reasons, the skipper is free to find another. An example might be cited.

One fisherman had for many years been a member of the core crew of a skipper, but he dropped out in 1973 due to disagreements over a loan. He had got a job in a construction company until in the spring of 1975 he again returned to the fishing business, now as a core crew on a boat of which the skipper was desperately in

Ill.27 A fisherman filling boxes of fish with ice. The man in the background is employed at the cooperative and records the quantities of fish.

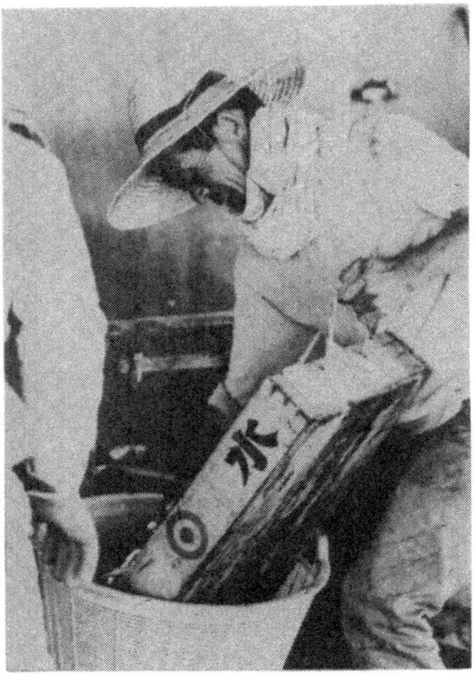

Ill.28 Fish being scaled before being put into boxes of six kilograms.

need of men. He worked for this skipper a couple of months until he was wounded in a finger by a poisonous fish. He could not fish for some time, but when he still did not return to the boat when he recovered, the other fishermen urged the skipper to drop him completely. The skipper was in serious trouble as the profitable *taichigyo* season was approaching, and the rules for this common endeavour state that there should be three persons on each boat. He managed in the end to recruit enough persons because of old ties with former members of his core crew who at that moment were fishing on their own vessels. Thus, the ties between the skipper and the core crew members are strong and can be utilized many years after the relationship on the boat has been dissolved. On the other hand, breaks in these obligations can be sanctioned; not only will the person lose the possibility of work on that particular boat, he will also have serious trouble in getting a job on another boat for some time. The social relations are therefore an important resource in the transactions between the skipper and the crew, and this will be discussed in a later section of this chapter.

The member of the core crew should know his own work. Though the skipper is the responsible person on board, everybody has to know what to do in order to fish efficiently. As the crew is small, the work is diversified. He has to be able to perform all the tasks, perhaps with the exclusion of steering the boat. Where the member is young and inexperienced, he has to possess the ability to comprehend the orders given, and perform the tasks without instructions as quickly as possible.

The core crew member has also obligations on land when the fish has to be sorted. This task is performed by the crew while the skipper takes care of the boat. The fish is sorted both according to kind and size, and the prices vary accordingly. A better price for the fish can be obtained if the right strategy is used during sorting. The fish is then scaled and loaded in boxes with ice and stored till the next morning when it is brought to the market in Hakata by an employee at the fishing cooperative.[3]

In return for his work the member of the core crew is entitled to a share of the net profit, and this is as mentioned equal to the skipper's. In a sense he also has the right to a certain degree of social security; the skipper will keep him for an indefinite time and will not let him off due to more mechanized equipment. He has no right, however, to be employed during the slack winter season. In time of stress he can nonetheless count on economic assistance from the skipper in the form of advantageous loans.

The Status as Member of the Extra Crew

Extra crew members are taken on board only when it is absolutely necessary. Today this is done mostly during the short *taichigyo* season when ideally all the boats should have a crew of three. A few boats have trouble in getting a third member, and occasionally people from outside have been hired. There are moral sanctions against operating with only two, as one informant told me: "Last year I had to hire a student from Fukuoka. Maybe it would have been possible to participate with a crew of only two, but it would have been embarrassing. I would have been regarded as mean and greedy. As a member of the cooperative committee I have to take care of my reputation." In the 1975 season this rule was nevertheless adjusted to allow one boat to participate with only two men. This was done because the boat had been plagued with bad luck and illness. Considerations will be taken in case of need.

Previously the extra crew was hired during the season for sand lance in the spring. This was the season when four to five persons were required on the boat as compared to a core crew (including the skipper) of two or three. Many persons were hired from outside the village, and they came even from other counties. Many farmers participated since it was still too early to start the cultivation of rice. This situation has drastically changed. The agriculture in the area has been intensified, and most fields have two crops: rice in the summer and vegetables or strawberry under vinyl greenhouses during the winter. The population engaged in farming has at the same time decreased, mostly because the husbands have taken work elsewhere while the wives and grandparents take care of the farming, a pattern called *sanchan nōgyō*. Farmers are no longer available as extra crew members, and the skipper might have difficulties in recruiting an extra member even in cases of emergencies such as illness.

The status as a member of the extra crew is often defined by contract, though not necessarily written. Unlike the core crew that can be called on at any time, the extra crew has the duty to work for the skipper on certain specified tasks during an agreed period of time.[4] When they do participate on the boat, however, they can be put to any kind of work, and they will be expected to sort the fish afterwards. Knowledge of the fishing operation is not necessarily required of him, but it will be

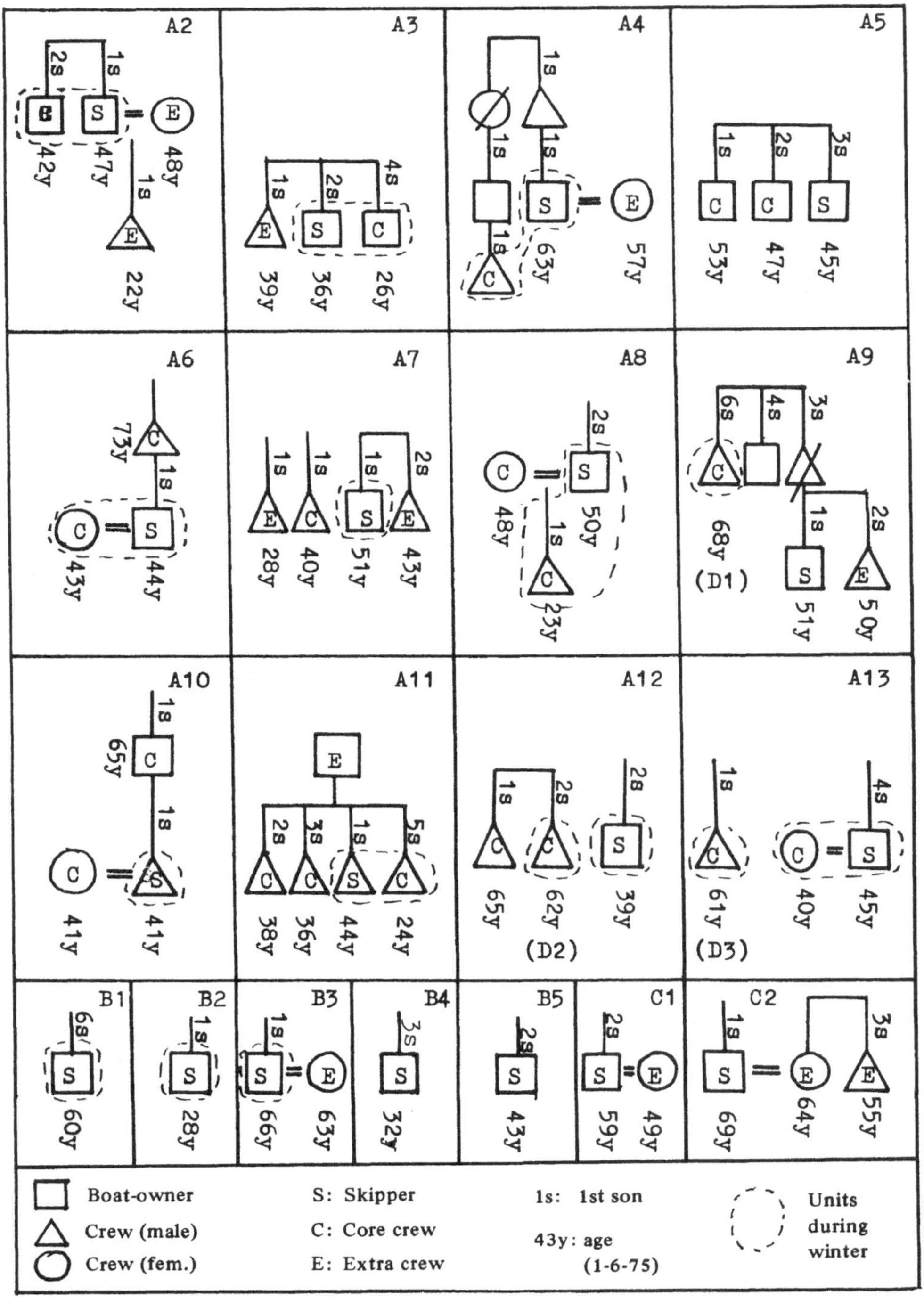

Fig 1 Composition of the crews as they were found during May-June 1975. A2-A13 operated *gochiami*, B1-C2 smaller nets and traps.

valued if he possesses this resource. An inexperienced man from outside Shingū or a youngster from within will not receive a full share, but at the moment 0.8 of this. It should be added that the status "experienced" is not an absolute, but a relative term. His closeness to the skipper influences this definition, but during the common endeavour this status will be defined at a meeting, where it will be decided if anybody should have reduced shares.

Compositon of the Crew

I have so far tried to outline the resources which are needed in order to hold the three statuses that I regard as crucial on a fishing vessel in Shingū. I also indicated that crew members might hold additional resources that would influence their role performance. As these resources as well as the role performances can best be understood in relation to the composition of the crew, I will use the latter as the point of departure and hence scrutinize the crews as I found them in June 1975. I have in Fig.1 indicated who were skippers, core crew and extra crew members; furthermore who held the status as owners of the boats (and nets). Other relevant factors are age, sex, and seniority in birth in a group of male siblings.

If we take a look at the boats *A2* to *A13* on Fig.1 (these are the boats that operate *gochiami*) some features stand out sharply. Most of the crews are based on either a group of male siblings (on six boats), the Father-Son relationship (on five boats) or a combination of both (on two boats). This is not a new trend; most of my informants said they started to fish with their fathers unless these died at a young age. This can lead to an impression that agnatic lineages are important in the formation of the crew. There are only three crews (*A4, A12* and *A13*) where no such relationship exists between the skipper and at least one crew member. On the smaller management units (*B1* through *C2*) no such relationships can be found, but it should be mentioned that all these skippers started their careers with their fathers or brothers. Their fishing operations today are, however, too limited to employ two men on a full-time basis. If we analyse the dyadic relations between the skipper and his crew, we find that there are ten cases of Brother-Brother relationship and five cases of Father-Son out of a total of twenty-eight dyads (Table 16).

Table 16 Dyadic Relationships between the Skipper and his Crew (1975)

	Gochiami boats (*A2-A13*)			Boats *B1-C2*
Relationship	core	extra	total	extra
Brother-Brother	7	3	10	0
Father-Son	3	2	5	0
Husband-Wife	4	2	6	3
other relatives	2	0	2	1
not related	4	1	5	0
Total	20	8	28	4

Ill.29 A fisherwoman. A recent development in Shingū has been for the skippers to engage their wives on the boats.

The Husband-Wife relationship is also important in crew formation; it is outnumbered only by Brother-Brother relationships. This is a trend that developed only recently, and it would have been unthinkable only a few years ago when the taboos surrounding women were still strong. The women were not allowed to enter the fishing boats, and they were only permitted to stand on the beach hauling in the beach seines and sea bream beach nets. During the war, however, at least one woman started to fish in the absence of her husband, but the wives were admitted to the *gochiami* boats only about 1970, and still they are restricted during the menstrual periods.

Two principles for crew formations have so far been examined. We have seen that the crews have been based on agnatic dyads and more recently on married couples. A third category of relatives can also be activated to recruit crew members, namely the *shinrui* that includes both cognatic and affinal relatives (Chapter IV). Both *A4* and *C2* are examples where such dyads to the skipper exist. Judging from the fishermen's life-stories as they were told me, this pattern was more prevalent in the past, probably because the crews were larger. Moreover, during the Tokugawa period the mean household size was about four climbing to five, which implies small families and few sons to take over the fishing enterprises. Cooperation from *shinrui* would in that situation carry more weight than it does today.

Affection as the decisive factor for crew formation seems to be of lesser importance. Boats have from time to time been owned and operated by friends, usually in the same age group (*dōnen)*, though there are no such cases at the moment. This kind of arrangement seems to be unstable as the affection cannot easily be transferred to their respective sons. An affection that exists between a patron and his client can also be utilized to form the crew, and one such crew has been formed recently (see Case 2 below).

Five cases of skipper-crewmen dyads on three boats (*A 7*, *A12* and *A13*) did not have any basis in family relations. These five were hired and did not possess other resources than required to perform the role as crew members. Though quantitative data is difficult to obtain on this issue, it is my impression that such dyads were more frequent in the past. I will in Chapter IX discuss the pattern of recruitment at length, but here it will suffice to say that sons of hired core crew members have not entered the fishing boats. All the five persons hired in 1975 were sons of boat-owners. Two of them operated their own boats during the winter season (*A12-C62* and *A13-C61*) and a third operated his own boat most of the year (*A 7-E28*).[5] The occupational mobility among the sons of hired crew members has restricted the development of social stratification among the fishermen and caused a labour shortage within the fishing industry in Shingū. This leads again to the need of incentives to attract the crew members and the equal division of the net earnings must be seen in this light.

Development Cycle of the Crew

Inheritance was, until the end of World War II, through the male line only, and primogeniture was preferably practised. The eldest son should inherit all the property, and it was his duty to take over the house and care for the aging parents. The other sons were – if the family could afford to do so – helped to establish themselves in independent branch households (*bunke*).

Today all the children have the same rights to the inheritance of their parents. In practice this means that at least all the sons should get equal shares of their father's belongings, though the contrary usually can be seen in many farming households where division would be destructive due to the small holdings. This rule does not apply to the inheritance of the shares in the fishing cooperative either. The full membership in the cooperative and thus the right to be a skipper cannot be divided, and it is given to the son who succeeds as the head of the household. If he does not fish but a brother of his does, the share might be given to the latter. On the other hand, other sons can acquire these rights to fish by buying shares.

Where several brothers own boats together, most of them are able to act as skippers. This is the case on the boats *A2*, *A3*, and *A11* where the persons *A2-C42*, *A3-C26* and *A11-C36* can serve as skippers in their absence. All the brothers on boat *A5*, which is a special case, can act as skipper. The official skipper (and the registered owner) is the eldest of the three, but the youngest is regarded as the most able of them and the *de facto* skipper. He has the longest experience, as both his brothers have worked outside the community. The position as skipper is thus not only determined by age, but also by experience and skill among the owners. Once the younger brother becomes the skipper, his position will not change even if the elder brothers subsequently acquire the same experience and skill. The younger brother remains the senior on the boat, and the situation can be compared with the seniority system in a Japanese company.

If all the sons begin to fish with their father, they will in theory get the same education and experience in the fishing operations. The eldest son will usually be the first to start fishing, and consequently he will also acquire the necessary skills before his younger brothers. The high value put on the younger's obedience to the older, which is embedded in Confucianism, is thus reinforced on the boat. But the younger will eventually get the same experience, and the resources he possesses will basically be the same as the skipper's. The stress on the value of seniority is not always enough to keep the younger brothers in line, and tensions do frequently arise. The critical phase of this development occurs when their father, who had inhibited the competing ambitions of his sons, retires. Two boats were in this critical phase during the time of my field-work. The conflict was obvious on boat *A11*, and the two brothers *A11-S* and *A11-C36* were observed fighting over the helm. The younger blames the elder because the skipper takes all the decisions without consultations. He feels that he has the right, by possessing the same resources as his elder brother, to participate in these decisions. In other words, he wants an egalitarian crew. On the other hand he also blames his brother for not being responsible. The skipper is known as a gambler and does not, so the younger says, take care of the equipment. The skipper on his part regards the privilege of avoiding some of the menial tasks as his inevitable right as the elder brother. And as the elder brother he feels he has the right and the obligation to take the leadership according to the high value put on the seniority of the eldest son embedded in the old civil code.

Another factor complicated the matter further; the skipper had established himself in a separate household making the succession to the head of the original house undecided. When I visited Shingū again in 1978 the two brothers had exchanged residences, thus clarifying the problem of succession to the headship of their natal household. Moreover, one of the other brothers (*A11-C38*) had retired because of poor health, leaving only three brothers on the boat. The youngest of the three strongly supports the skipper, and the tensions seem to have cooled down a bit.

There is still no indication of problems on boat *A3* where the father of the brothers died a fortnight after our first arrival in Shingū The eldest of the three brothers only fishes when an extra crew member is needed, usually he is a taxi-driver. The youngest is on the other hand still young and in the learning process, and he does not yet possess the resources to challenge the leadership of his elder brother who succeeded to the headship of the household on the death of their father. One further note should be made. The core crew of this boat consists of only two men, which is regarded as the minimum required to continue their technology. A fission at the moment would be destructive to both.

The lack of potential crew members is in my opinion one of the most serious restraints on fission. Brothers will be forced to continue together until more members are available. Usually this possibility opens when their sons graduate from school. This is the second critical phase.

Two cases of fission will be discussed in some detail. The first was successful, the second was not.

Case 1: The persons *A13-S* and *A8-S* (Fig.1) are brothers and used to work together under their father who was one of the most prominent fishermen at that time. The eldest of the sons (*A8-S*) became the new skipper and his younger brother remained member of the core crew. The father became extra crew member and a cousin was the third member of the core crew. When the son of the skipper started to work on the boat, the younger brother saw an opportunity to establish himself, and he received the old boat as financial assistance from his parental house (i.e. his *honke*). The boat of his elder brother had a sufficient crew; his problem became to gather one for himself. Allegedly he expected to mobilize a second-cousin (who was *bunke* to his father's house), but was refused. He then turned to his present core member *A13-C61* who previously had been working many years for *A13-S*'s father. He was available at that moment since the boat he used to work on (*A1*) had just ceased operations. *A13-S* also succeeded in hiring a third member *B4-S*, and his crew was secured. Later *B4-S* dropped out to take over his father's boat, and the skipper had to bring in his wife in his place. A similar development occurred on the boat *A8* where the cousin disappeared from the community. Father and son continued alone until the latter hurt his back. The wife started to help and from 1975 she has been a member of his core crew.

Both brothers have had difficulties in securing their crews since the fission, though the fission seems to have occurred on a favourable time when crew members were available. Nonetheless, they have at present stable crews and their catches are reasonably good, the youngest has in fact one of the two very best results.

Case 2: The factor that released the fission described above was said to be disagreement over economic matters between the wives of the two brothers. The same was said to be involved in the second case. The two brothers are *A7-S* and *A7-E43 alias B5-S* in Fig.1. The younger brother obtained after the fission only a boat in poor condition unsuitable for *gochiami*. He began therefore to fish alone using smaller nets and traps. The elder brother continued with one of his sons, who disliked the work and dropped out after one year. Another member of his core crew (*A7-E28 alias B2-S*) started to fish alone when he inherited his father's old boat. The skipper became desperately in need of a crewman and made an agreement with *A7-C* who, however, wounded his finger soon afterwards. *B2-S* rejoined the boat due to old obligations, and the brother of the skipper joined as extra crew during the *taichigyo* fishery. Both continued after the closure of this fishing, and *A7-E28* explained his position thus: "It is impossible to participate only during the very profitable *taichigyo* fisheries and then just disappear. I felt obliged to continue also when the fishing was not that good. So, next year I will probably not participate at all to avoid later obligations." He had kept his word.

The situation was untenable in the long run, and the two brothers agreed to work on the same vessel again. They managed to get *B4-S*, a distant relative (both a *shinrui* and member of the same weak *dōzoku*) as the third member. The ties between the skipper and *B4-S* have long been intimate, and the skipper has taken a special interest in the young man. He was the obvious go-between in the latter's wedding in 1975. These three men bought a new boat together, and *A7-S* remains the skipper on the boat that started operations in the spring of 1976.

In this last case the two brothers merged only when no non-agnates were available. From Newfoundland Nemec reports that the fishermen there fish together with non-agnates only when they lacked agnates (Nemec 1972:32), and a similar situation is reported from Sweden (Lofgren 1972:93). In Shingū the situation seems in some ways to be reversed; they fish with their brothers only when there are no other possibilities of forming independent crews with themselves as skippers. Due to the decrease in the number of fishermen, a process that is particularly strong among the hired crew members, brothers are forced to extend the period of co-

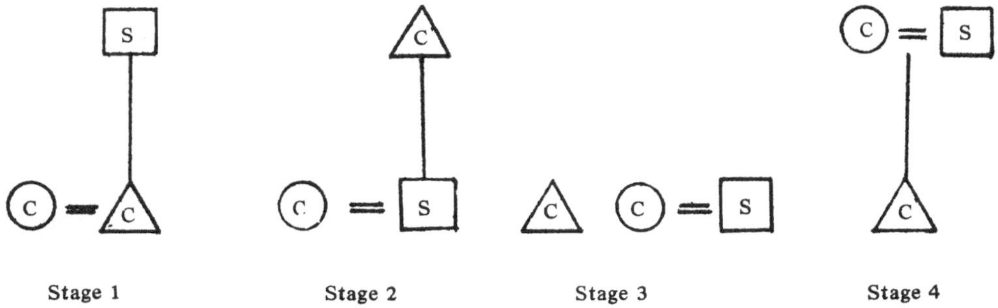

Fig.2 A tentative development cycle of a crew.

operation, which leads to more egalitarian crews since they possess the same resources.

The brothers in Case 2 failed to gather members for their crews, while the brothers in the first case managed this in the end only by bringing in their wives. As we have seen, this is a new trend, and if we allow for this possibility, we may tentatively look upon the crews on the boats *A6, A8, A10* and *A13* as stages in a process similar to the developmental cycle in the household.

Boat *A10* had in 1975 just passed Stage 1 as the son took over as skipper, but he was still not the owner of the boat. This he did a couple of years later, and in 1978 the boat had reached Stage 2. Boat *A6* has been in the second stage for several years and is moving towards the critical Stage 3. The married couple on the boat are prepared to continue alone when the father retires, but an additional crew member might be hired – boat *A13* is an example of this development. Finally the son of the couple may take his place on board, and boat *A8* is in this stage (Stage 4).

Let me stress that this is a hypothetical cycle; I only want to explore the potentials in the new trend bringing the wives to the fishing grounds. It all occurred too recently, and only boat *A10* has passed from one stage to the next. But the possibilities are evident, and will probably be utilized when needs arise. Hence, several of the skippers did express the advantages of bringing in their wives instead of hiring people as core or extra crew (cf. below), and a fifth skipper (*A12-S*) had by 1978 engaged his wife.

The women have the same obligations on the boat as the male core member. Their rights are on the other hand restricted. Firstly, they are certainly not allowed to change boats; if they fish they have to do so with their husbands. Secondly, they are at least at the moment excluded from the status as skipper. Thirdly, they are not entitled to the same share as male fishermen. Where the crew consists of one household, no distinction is made unless the boat is participating in a common endeavour. Their shares are then reduced to 0.8 whether they belong to the core or extra crew. The justification for this decision is that the women are allegedly not as strong as the men and thus contribute less to the catch. Fourthly, they are excluded from the decision-making at the fishing cooperative. As the household holds only one vote in the cooperative, the husband will control this.[6] He is allowed to send a deputy, but no woman has to my knowledge been sent in this position so far.[7] Fifthly, they are excluded from religious and recreational activities connected with the fisheries. They are not allowed to participate during the *Ryūgū* festival, nor on the annual excursion of the cooperative. In many ways the woman can best be compared to the male members of the extra crew. Her rights are limited, but her obligations are those of the core-crew status.

Final Comments on the Role Performance of the Crew

Some comments on the role performance have already been made in the discussion above, but some final remarks will be added. The composition of the crew does, as we have seen, greatly influence its performance, and there are tendencies to fission where the skipper and the crewmen possess the same resources.

The most stable crew seems to be obtained when there is clear distinction in the statuses on the boat and the resources the personnel possesses; i.e. when the crew is a vertical structure with the skipper as the ultimate authority. The members will in this case do anything not to challenge this authority. Under-communicating their knowledge and ideas was a stratagem frequently used. Several stated that they would give an opinion only if directly asked, and that is unlikely. One crewman admitted that it was very frustrating not to be able to voice an opinion; even the best skipper could sometimes make mistakes. If the crew was composed of a single household, it would have been easier, he continued, thus indicating how social relations do modify the role performance of the crew.

The crewmen are – where the vertical structure persists – members of the crew only through their dyadic relations with the skipper, not as co-owners of boat and

Ill.30 A fish pedlar. This has traditionally been the work of fishermen's wives and widows. In Shingū there were four such pedlars in 1975.

equipment. The starting mechanism of the transaction between the two was traditionally a mutual need of the other's resources. The skipper needed labour and the crew member a place to work. The relationship between the skipper and his crew does usually extend beyond the point of formal contract unless it involves a member of the extra crew. The dyadic relationship between the skipper and his core crew member will often resemble the relationship between an *oyabun* and *kobun* which is often translated as the patron-client relationship (Nakane 1967, 1973, Fukutake 1972), though Bennett and Ishino argue that the patron-client concept "expresses the hierarchy but misses the simulated kinship element" (Bennett & Ishino 1963:30-31).

The skipper tries to establish close personal ties with his crewmen in order to secure a stable working force. He extends economic aid to his men in time of need, he may act as negotiator and go-between in their marriage; he should in any way behave "kindly" towards his crew members. One complained about failure in this respect in the following terms:

> The trouble started when the skipper took his wife on board. It was much better when X worked; he could really work as a man. But that's not the main point, you see. It's problems with women on board. If you employ anybody, you ought to treat them nicely and be kind towards them. But she uses rough speech and shows no respect. Before, when I used to fish with Y and we got tasty fish, he always told me to keep some of the good fish for myself. This never happens to me now; she doesn't show any consideration.

This man obviously found some difficulties in his position on the boat, and he felt the position of the skipper's wife ambivalent. On the one hand she was a member of the core crew and should behave as such, on the other hand she could utilize her personal relation with the skipper in performing her role. He expected her to perform accordingly; i.e. in a similar manner as the skipper, his *oyabun.* She ought to show respect, kindness and consideration; she did not realize that a "minimal charge of affect invests the relation of patron and client" (Wolf 1966: 16). This crewman did not feel any deep loyalty to his skipper, and he explained that he continued fishing only because the skipper was very successful. If there are three consecutive days of poor catches, he allegedly will not continue. He would then fail to extend to his skipper the expected services of loyalty and the feeling of gratitude. In Bailey's terms he has ceased to extend long-term credit and has become rather a follower than a core member (Bailey 1969).

The resources of a skipper are limited. He is neither rich nor does he have a stable income. His abilities to act as *oyabun* are therefore seriously restricted. New limitations have been added as the possibilities to obtain work outside the fisheries have improved. The skipper has to increase his investment in the relation with a crew member in order to establish himself as a patron, and he might neither be able nor willing to do so. The costs and the rewards are no longer commensurable.

The skipper has today few possibilities for sanctioning undesired behaviour of his core crew. It is difficult to get a full crew, and several of the skippers complained about the lack of trust and loyalty. If the fishing was good, the crew would surely show up, but sometimes the catches are down, and the crew might fail to come. The skippers have difficulties in handling this uncertainty which restricts their ability to make sound plans for the day's operations. They have few opportunities to exercise the authority embedded in the *oyabun-kobun* relationship. This is particularly so when the crewman is not the sole money earner in his household. With this background it is easy to understand why the skippers may choose to engage their wives, and two skippers claimed that the troubles mentioned above were the reasons why they decided to do so. The economic considerations came second.

It becomes an important issue for the skipper to be successful as a fisherman in order to keep his crew. The absolute value of the catch is of course decisive to the viability of his crew. If the income is low, members might disappear. But the catch is also valued in relative terms. A good catch compared to the others improves the skipper's prestige. The crew might be utterly depressed after a miserable day, but then be happily relieved when they discover that the other boats got even less. The crews checked their sale-rating from sale-lists kept at the harbour every day, and they were aware of their position *vis-à-vis* the others. And one skipper was always met by his daughters asking his rating, not the value of the catch. The high attention given to relative catch seems to be quite universal (Andersen 1972, Barth 1966:9-10), but this attitude extends also to the boats with family crews. This phenomenon can therefore not simply be explained in terms of the competition for good crew members. A high rating gives prestige which in certain environments can be converted to political influence. This was the case in Shingū in the past, but less so at present. The prize today is prestige, honour and self-esteem, and the value of the prize is regarded so high that it is worth fighting for even when the crew is stable as on a family boat.

With this highly competitive attitude it is understandable that everybody stays at home on the same days. The weather conditions are of course often such that all agree that it is terrible. But the weather has also to be seen in connection with expected catch; if the results have been bad for a couple of days, they feel discouraged and define a little wind as "bad weather". Or, on the contrary, if the fishing has been rich, they will defy the waves. (The situation is similar to that reported by Firth (1966) in Malaysia.) The point is that everybody seems to be agreed in this respect. The same holds true when the reason for staying home is "tiredness" or "low prices".

Usually the fisherman scales the sand dunes early in the morning to take a look at the sea. If the conditions are found agreeable, he goes home to eat breakfast and prepares a lunch-box, and then proceeds to the harbour. If he is in doubt, however, he will again scale the sand dunes and wait there. If the next fisherman also stops, this indicates that he is uncertain too, and they discuss and wait. If everybody stops, they will probably decide that the weather is too bad or that they are too tired. If one of them continues straight to the harbour, the others will follow. Once eleven of the twelve skippers on the *gochiami* boats had gathered at the harbour and decided not to fish. The twelfth skipper who came a little later, did not pay any notice to the others, jumped into his boat and prepared his nets. The others changed their minds.

To summarize: I have in this chapter tried to show how the personnel on the boats could be analysed in terms of three statuses. Certain resources were needed to qualify for each. Where the personnel possessed different resources, a hierarchical crew developed, which the skipper tried to make stable by knitting the core crew members to himself in patron-client relations. With the industrialization, however, new attractive work-opportunities opened to the fishermen, and the skippers had to invest more in the relationships with their crews in order to keep them. The crewmen have managed to get into a bargain position and the skippers have few opportunities for successful sanctions. Today it is the skipper who needs the crew, and it is primarily by using another relationship of authority (as between Husband-Wife and agnates) that he sometimes manages to press others to join him on the boat. This entails profound modifications of the traditional vertical pattern. The crewmen have come to possess more or less the same resources as the skipper. The crews have become more egalitarian and the skippers have become "weaker" leaders.

CHAPTER VII

NET-GROUPS AS MANAGEMENT UNITS

It has been mentioned that there existed structures where several boats cooperated in different ways. These are the net-groups. By a net-group (*amigumi*) I will mean a group of at least two crews that cooperate by operating nets jointly. The equipment can either be owned individually or jointly by the fishermen themselves, or by a capitalist who does not take part in the fishing operations. It will be obvious that this definition of net-groups covers a wide range of different organization forms. I will therefore begin this chapter with a description of the more important net-groups found in Shingū today and in the past.

There has been a wide variety of net-groups through the ages; from small groups comprising only a couple of boats and a few men to large structures with fifty members or more. The behaviour of the fish and the nature of the technology used to catch them demanded certain requisites of members and organization of groups, but often ownership and kinship influenced the composition and relationship within these groups. The net-groups can, according to ownership, be classified into three distinct types. The first is based on an owner (*animoto*) who engages fishermen to operate his equipment, the second is based on a kinship group, while the third operates cooperatively owned nets.

A few of these structures were formed by cooperation between some of the boats analysed in the previous chapter. The status sets found on these boats were therefore in some cases transferred to the bigger management units, though a fourth status was introduced; the status of a leader or co-ordinator. When the group operated jointly owned boats and nets, however, completely new status-sets developed. The ownership of the equipment used seems to have been a crucial factor in the operation of these groups, and attention will be given to this aspect in the second half of this chapter where I will focus on the leadership of the net-groups.

Joint Marketing Cooperatives

This kind of coooperation is not strictly speaking a net-group as defined above. There is no joint operation of nets on the fishing grounds. This kind of cooperation is nonetheless extremely important in Shingū today, and it also offers a convenient link between the individual fishing described in the previous chapter and the net-groups operating common nets and boats. The important characteristic of the cooperation is the joint marketing, but some equipment might be owned by the cooperative.

Though joint marketing cooperatives achieved their importance only recently, they have also been used occasionally in the past. A net-group named Kanmata operated the sand lance fisheries in this way. They operated two independent boats each with nets, but pooled their catches and divided the income.

The importance of this kind of cooperation became evident with the emergence of the *taichigyo* fisheries. Each boat is fishing independently on the fishing grounds. Hence, this kind of fishing could easily have been conducted on an individual basis.

In fact the fishermen in the neighbouring community Nata are doing so. Also the Shingū fishermen started individually in 1957 when Shingū became the pioneering community regarding this fishery in the area. The first trials were far from successful, and half the fish died. The Shingū fishermen became known as the "fish killers", and it was a painful and demanding process to learn to keep the fry alive. In this they got invaluable help from a veterinarian who was an expert on fish breeding, and today the losses are almost nil. The fishermen are rightly proud of their *taichigyo* which is known as among the very best in quality, and the prices are correspondingly high.

After a few years the fishermen began to cooperate by owning some of the equipment in the harbour together, and finally by pooling their catches. The fishing cooperative was actively promoting this cooperation, and it was the only way that everybody could benefit fully from the veterinarian's knowledge. During the first years of cooperative efforts quarrels often erupted at the harbour, and the fishermen violently accused each other for being lazy and ignorant, and fights occurred.

All the *gochiami* boats take part in this fishery today, and several crews switched to *gochiami* in order to participate in this profitable fishery. The profits have been spectacular; during the best years more than half the income of the year has been derived from this fishery during a short summer month. The fishery reached its present peak in 1977 when each crew member received 1.5 million yen.

The boats leave the harbour as usual in the morning and return about 5 p.m., all at approximately the same time in contrast to the ordinary *kisugo* fishery when they arrive with a couple of hours between the first and the last. The fry is carefully loaded into twenty-two nets that are kept in the harbour. Other species are then sorted jointly and sold as usual at the market in Hakata the next morning. All the fish belong to this cooperation. If a fisherman wants fish for dinner (something he always wants), he has to pay for it. The prices for each kind of fish are decided at a meeting. Occasionally fish is given to everybody, but this can only be practised when sufficient quantity of one species is caught enabling a division into thirty-nine parts; one each for the participants and for the three employed at the fishing cooperative. The valuable *kisugo* was sometimes distributed in this way, a fish that was very rarely taken home when fishing individually. Less valuable fish, which they usually consume themselves, are caught in smaller quantities and are brought to the market or minced to be fed to the fry. The income from fish brought to the market is spent to cover common expenses and for an excursion at the end of the year.

The income from the fry is distributed according to rules decided at a meeting where all involved may participate, though no woman has so far ever met. Each experienced male fisherman gets one share, women and extra crew members who are not experienced and junior high school students get 0.8 and children from elementary school (I have never seen nor heard of any participating) get 0.6 share each. The boat counts as 0.5 share. A boat with two men and one woman is thus calculated as having 3.3 shares.

An example will be given to illustrate the distribution of the income. The total number of fish sold is divided by the total number of shares and then multiplied by the number of shares for each boat to find out how many fish each boat contributed to the sale. In this case, one boat with 3.3 shares received:

11,665 *madai* sold for 90 yen each	= 1,050,000 yen
2,800 *chidai* sold for 40 yen each	= 112,000 yen
Total gross income for this boat	= 1,162,000 yen

From this gross income ten per cent was subtracted for boat-savings, ten per cent for equipment, six per cent to the fishing cooperative for marketing, 2000 yen/day for petrol (the above income was calculated as six days) and 1000 yen/day for excursion money. Net income was then 841,880 yen which was divided by 3.3 giving 255,110 yen for the men, 204,088 for the women and 127,555 yen for the boat.[1]

There is little cooperation on the fishing grounds, though everybody benefits from good catches. Radio telephones were installed in all the twelve boats in order to facilitate communications between them, but most are broken and no effort is made to repair them. Some information and advice are exchanged, however, in contrast to the negative sharing of information usually prevalent. On the shore some minimum of organization is necessary, and the twelve boats are divided into three sub-groups each with an elected leader. The three leaders are elected annually at a meeting. In theory all the participants are eligible except the skippers and crew members on boats that were represented the preceding year. In practice only skippers are elected in almost all the cases. As the leaders have to be responsible for each of the three sub-groups, only one from each boat can be elected. If two persons from the same boat are elected, one will be dropped and the person with the fourth highest vote will take his place. In the election in December 1975 nine persons received votes, all of them skippers. It is also remarkable that these represented all the nine boats eligible. If an elected leader cannot participate during the fishery, another man from the same boat will replace him. This was the case in 1975 when *A5-C47* was the leader in place of *A5-S* who was ill.

When the leaders have been elected, the composition of the sub-groups is decided by drawing lots to avoid favouritism and the development of cliques. It is the leaders' responsibility to ensure that the cooperation on land goes smoothly, that people are available for sorting the fish for the market etc. They have to decide when the fry is to be counted, when fish shall be distributed to everybody free of charge, and one of them has to keep accounts over the sale of fish to individual fishermen. They are also held responsible for the fry being fed.

The fish are fed on a rotational basis, one sub-group does the job each day, and the fish are fed morning and evening. Each boat in the sub-group should contribute one worker for this task, though this is not always the case. Once I registered that one man had to do the job alone, he was also the leader of the sub-group; one forgot it was his turn, and two came too late from betting on a boat-race. As the task is demanding, the elder men and women are exempted from feeding. When the fish have to be counted (a procedure that is needed because the individual catches are kept secret to avoid bad feelings and rivalry between the crews), the women and old men are also exempted. Finally it is the leaders' job to decide whether to fish or

Ill.31 The *taichigyo* fishery. At the end of the day the living fry are carefully transferred from the boats to twenty-two nets kept in the harbour.

not, but as the profits are high, the days of "bad weather" are few and far between. Dead days are nevertheless needed to count the fish, and it is significant that the fishermen take this loss to avoid conflicts that would erupt if the individual catches were made public.

It should be stressed that the status-sets found on the boats are preserved from the individual fishing. The status as leader does not permit anybody to interfere with another skipper's operations on the sea. Furthermore, all the negotiations with the authorities are made through the fishing cooperative, either by the manager or by the elected committee. The influence of the leaders is therefore limited, and no economic compensation is given.

Amimoto-Based Net-Groups

From the English literature on Japan, little is known about centralized ownership of nets, though both Norbeck (1954) and Brameld (1968) briefly mention *amimoto,* but they do not elaborate.

The *amimoto* existed formerly in Shingū and in some respects dominated the lives of the fishermen during the Tokugawa period and Meiji era. During the period 1823-1831 loans were given from the *tame-gin* fund at least six times for repair of nets. During the same period loans were given at least thirty-nine times to finance boats.[2] These numbers indicate the dualism in Shingū's fisheries; the independence of the fishermen in some fisheries illustrated by the high number of privately-owned boats on the one hand, on the other the existence of a few large nets. Unfortunately, it is difficult to know anything about the social background of those who borrowed money to finance the nets. One sea bream net was, however, owned by three fishermen from Honmachi. The owners of the other nets were probably from the merchant class. Two kinds of *amimoto* can thus be identified:

1: The first I will call an *absentee amimoto,* i.e. a person who is not engaged in fishing himself. Usually he is one of the leading merchants, money-lenders or the village headman. (These roles were often combined in one person as was the case with Kaneuchi, Yayama and Toyomura.)

2: The second is an *active amimoto;* the net is owned and controlled by an active successful fisherman. His economy is not as secure as the economy of the merchants who managed to found their financial "empires" on many assets. A successful *active amimoto* was nevertheless able to convert his success as a fisherman to influence in village affairs, and a few managed to buy some farm land.

It has been assumed that big landlords developed due to the introduction of money and the market institution during the Tokugawa period, but Smith (1952, 1959) has argued that there were marked distinctions, economical, political and judicial, between landholders and the others already before this period. The taxes were levied on land, and only the landowners had to pay these. In return they had several privileges. Those with no land had neither duties nor rights except towards the landlords for whom they worked. They had no rights to participate in the village affairs. The landholders were again a heterogeneous group; from rich landlords who controlled the village down to the smallest holder with land insufficient for survival.

So far the farming communities. Our question now is: did a similar pattern exist in the fishing villages? Yamaoka in his account from Waki-ura in Shimane Prefecture writes that the village headman Ōgawa owned many nets and controlled fishing territories exclusively. Fishermen who did not work for him were not allowed to fish inside these territories, but had to rent areas from adjacent villages (Yamaoka 1974:278-283).

Ōgawa operated as a typical *absentee amimoto.* He owned the nets, controlled the fishing rights and had the power to force the fishermen to work for him. In return he extended security and protection to his fishermen, and he was a prototype of an *oyabun* or patron.

The Kaneuchis might have played a similar role in Shingū, but it is doubtful if any single merchant ever developed complete monopoly as Ōgawa did. The position as village headman changed frequently, and there were serious economic fluctuations in Shingū, at least during the second half of the Tokugawa period. The house of Kaneuchi declined to give rise to Yayama that crumbled soon afterwards. At the end of the period Toyogaya and Yorosuya emerged as the leading houses, and both came to dominate net-groups.

There were several contrasts between Shingū as a fishing village and the typical farming community with its landlords and tenants. First, the economy was unstable due to fluctuating catches for the fishermen and the varying degrees of skill and luck on the part of the merchants. Big estates such as Yayama's could be scattered overnight. Secondly, the fishermen were far from self-sufficient. They were market oriented, and the fishing villages were often semi-urban in character. Thirdly, the Shingū fishermen had no land to mortgage except their own housing plots which were of limited value. In 1888 most of the fishermen owned their own houses as they do today. The biggest landowner at that time, Toyomura Heishirō, owned only one compound in Honmachi.

It is therefore doubtful whether a strong feudal society existed in Shingū. This does not mean that institutions like the *amimoto* did not exist, but he probably did not emerge as dominant as Ōgawa. His involvement was limited to certain fisheries where big capital investment was necessary, i.e. to finance certain nets, in particular the sardine nets. The sardine was often dried and used as fertilizer and was thus not dependent on the same distributional system as fresh fish. The sardine could be accumulated and traded in big quantities; it fitted better with the interests of the merchants.[3] Moreover, the taxes during the Tokugawa period were at least in part paid by money obtained from the sale of dried sardine. This income was supposed to be used to pay the *ura-unjōgin,* a fee levied on business.

As tax-paying was a shared responsibility of the whole village, everybody was morally obliged to participate in the sardine fisheries. Big and expensive nets were, however, necessary and the fishermen could not themselves raise the capital. Several merchants furnished loans to the fishermen or rented the equipment against security in future catches (Schepers 1935:59). The terms on which such assistance was given in Shingū are not known, but later in the Taishō era (1912-1926) the *amimoto* received about a third of the catch. They received this partly in kind, and thus obtained cheap and secure fertilizer for the fields they often possessed.

More research is needed to comprehend fully the organization of the net-groups in the Tokugawa period. During the Meiji era there was probably an increased economic participation by the fishermen. I was told that due to increased mobility after the Meiji Restoration, incentives had to be offered to the fishermen to keep them engaged in the net-groups, and many of the fishermen became minor co-owners.

In the beginning of this century two big net-groups based on *absentee amimoto* existed in Shingū. One was based on Toyomura's *sake* brewery (the group was called Toi-ami), and the other was based on Sakai's Yorosuya's *sake* brewery (Yorosu-ami). Both operated purse seines and lift nets *(hachidaami).* It seems that everybody participated in one of these net-groups. Sakai and Toyomura were not the sole owners; smaller share-holders among the affluent houses existed, and some of the fishermen also had shares. One manager who was one of the minor owners, was appointed by the share-holders. The fishing operations were led by a few skippers (*sendō*), and this was said to be an inherited status. Both the manager and the skippers got paid for their work, but it is not known how much they received.

Toi-ami and Yorosu-ami were the last net-groups that were based on *absentee amimoto,* and both were dissolved before World War II. The most important reason was allegedly the lack of profit on behalf of the big owners due to the absence of *maiwashi.* Moreover, Toyomura had by this time run into serious difficulties and left Shingū. The owner of the hospital, Araki, who was also from an ancient family of high standing, took over as *amimoto.* Some fishermen can remember that they got free hospitalization, and that they brought fish to Araki. But Araki ran into trouble too, and Toi-ami was dissolved.

The *absentee amimoto* was outplayed but the *active amimoto* continued for some more years. The net-groups that these founded were smaller and most of them were forgotten a long time ago. They were mostly short-lived and dependent on able leaders. The dyadic relationship between the leader and his employees could seldom be transferred to the next generation. In this respect they resembled the situation on the boats as described in the previous chapter. At least two such net-groups have existed in this century, and the sea bream net that was repaired in 1829 was probably owned in this way.

Inoue Sankichi, who was one of the influential fishermen from the 1930s to the 1950s, was operating sardine beach nets and later also two-boats *gochiami.* Sankichi, who was a member of one of the bigger groups, established his Sankichi-ami in the thirties. He also founded a processing industry where he dried the sardine his net-group could catch. After World War II Sankichi also bought fish from others to exploit the capacity during the winter months to avoid the work in construction companies. Sankichi himself fished and his wife administered the industry which employed from ten to twenty women during its best years. Sankichi-ami did none-theless have difficulties in hiring enough fishermen, and in the beginning of the sixties became merged with another net-group facing the same troubles. The division of the catch seems to have been typical for *amimoto* based net-groups; the owner got thirty per cent for nets and other equipment. About ten per cent was spent on running expenses like fuel, spare-parts and *sake* to the crew. The responsible person (in this case Sankichi himself) then received 1.2 shares, the employees one share each, the women pulling the nets 0.7 share each and one share for the boat (owned by Sankichi).

Kirishima Hangorō was one of the leading personalities in Shingū during the first twenty years of this century. In 1912 he became *kuchō* (ward headman), and he was later elected as representative both to the temple and shrine, as was Sankichi. Hangorō managed to operate a small beach net, but it is not known whether he used it to catch sardine. At any rate, he also participated in Toi-ami. Hangorō employed mostly relatives, and the core of his net-group was a small lineage extended by a *shinrui* relation as was Sankichi's core. This leads us to the second type of net-groups, those based on kinship groups.

Lineage-Based Net-groups

Though *dōzoku* does not seem to have been as strong in Shingū as in many other places in Japan, such kinship groups were nevertheless utilized to solve organizational problems relating to the fisheries. The most important of these kinship groups was the Matsuo lineage, that in 1975 included sixteen of the twenty-nine fishing households. Whether all these households really belong to the same patrilinear lineage is doubtful, however. Several of the households have other surnames; one got the name by a *pro forma* adoption to avoid conscription in the early Meiji era, a strategy no longer possible. The background of the other is more uncertain, though all have received adopted heirs from the Matsuos. In fact, the adopted son often refers to his original parental house as his *honke* thus ignoring the lineage connections of his new household. Households have in this way changed their *honke-bunke* affiliations, another indication of only weak *dōzoku*. At any rate, all the sixteen households agree on one household as the ultimate main house, though this household does not have any effective authority any longer.

Matsuo were among the first settlers in the present location of Shingū, and they came to dominate the western part of Honmachi. In Honmachi-5 seven out of eleven households bear the Matsuo surname.[4] The kinship ties have thus been strengthened through cooperation in local groups.

Originally the households in the Matsuo *dōzoku* formed a corporate group through their cooperation in a net-group named Kitagata-ami. It is not known when this net-group was founded, but it probably has its roots back in the Tokugawa period. During the Meiji and Taishō eras the members also participated in Toi-ami. About 1925 Kitagata-ami emerged as an important net-group operating large beach seines catching yellowtail and Spanish mackerel, and when Toi-ami was dissolved the group also operated sardine beach nets. Kitagata-ami operated a wide range of smaller nets as well, and few of its members ever fished by hook-and-line as was common among the other fishermen in Shingū at that time. Kitagata-ami was dissolved only in 1977, but it survived that long only through fission, fusion and re-organization.

If a *dōzoku* should be effective, there must be a stem house having resources to exercise authority. These resources are of two kinds: economic and ideological. Strong *dōzoku* developed where, as Nakane pointed out, wealth could be accumulated, and this was difficult in a fishing community like Shingū. The main household of the Matsuo *dōzoku* did not control the economic resources necessary to act efficiently as the leader, and disputes erupted. Whether the dispute started in the *dōzoku* itself or in the net-group is not known, but the effects became profound.

Ill.32 The ruins of a fish-drying house once operated by one of the large net-groups.

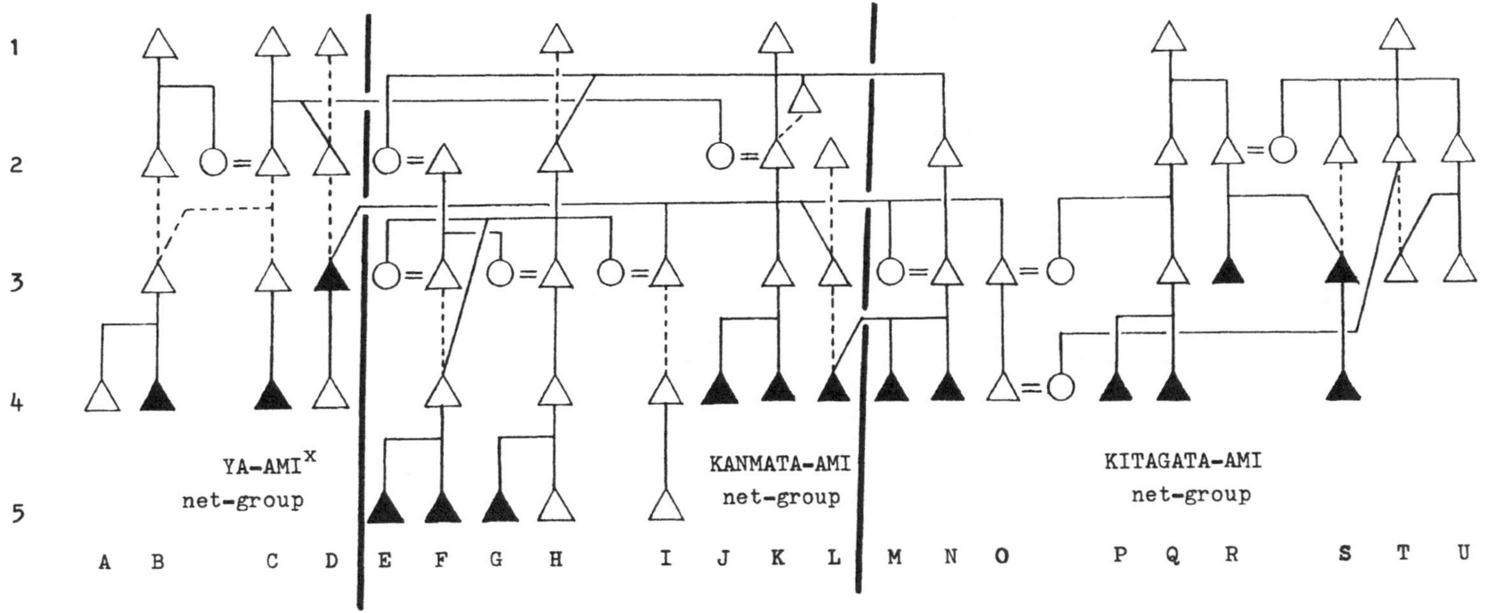

Fig.3 The Matsuo kinship group and its division into net-groups.

Notes: Not all branch households are indicated on this chart, but those which are omitted have all failed to continue in the fishing business. Dotted lines indicate adoptions; black triangles indicate active fishermen at present (1.1.1976).

x: This is not the real name of this group which operates small yellowtail nets; "YA" are the initial letters of B4 and C4's given names.

Everybody within the kinship group claims that *honke* is in line *K* (Fig. 3), but some people outside the group say that *Q* is in fact the real *honke*. Actually, *Q*, which by most people is referred to as *inkyo* (retirement house), has long been the richest and the most influential of the two houses. *Q2* donated twice as much money to the temple-repairs in 1908, which at that time was an improper act if *Q2* was *bunke* of *K2* in an effective *dōzoku*. (*Honke* ought to donate more than its branch households to show its superior position.) *Q2* was also active in the community affairs which *K2* was not. Allegedly both competed for the leadership in Kitagata-ami with *Q2* as the victor. When and how the fission of Kitagata-ami came about was a subject which was carefully avoided. Nobody knew anything, but all agreed that the conflict must have been severe.

I was given the most complete version of the story by the eldest member of the fishing community. He held high positions both in the cooperative (as chairman for many years) and in the township. According to him (and others) *F4* in Fig. 3 was a very hot-tempered young man who wanted to establish his own net-group, which my informant found reasonable. All young men with ambitions want to do so, he said. *F4* then talked to *I3* who was married to his sister. This was an act my informant regarded as a little improper as *I3* already participated in a net-group. *F4* had tremendous success; all the brothers of *I3* but one left Kitagata-ami, and so did his new household into which he was adopted. All, including his natal house joined the new Kanmata-ami; the term was composed of the first ideogram of the name of *F4*'s adopted father and the first ideogram of the name of I3's brother, the Matsuo *honke*. Kanmata-ami was, however, too small to operate its own sardine nets, and the group joined another net-group called Nigō-ami (to be discussed in the next section).

Kanmata-ami operated many small nets among which sand lance nets, two-boats *gochiami* and *nuikiriami* were the most important. Kitagata-ami sought cooperation first from Yayama-ami, which was another lineage based net-group centred around Yayama Gonji and later Yayama Rinzō, before it eventually merged with Sankichi-ami to secure working power for its sardine nets and *nuikiriami.*

Both Kitagata-ami and Kanmata-ami were too small to continue independently, and the fission can therefore not be understood in terms of a regular segmentation of a large kinship based net-group that had grown too big. The fission can only be understood in the light of the lack of authority of the main household, whether it was *K* or *Q*. The competition between these two houses and the personal enmity between *K2* and *Q2* paved the way for a break; the only thing needed was an entrepreneur of *F4*'s abilities to stage the fission.

The dispute had consequences not only for the net-group, but for the Matsuo *dōzoku* as such. *O3* was the eldest son of *K2* and was to succeed as the head of this important household. But he fell in love with *Q2's* daughter who lived next door. Their marriage was sanctioned by both parties. *Q2* became allegedly furious and expelled his daughter; *K2* disinherited his first son but built him a house in the backyard. A reunion between father and daughter occurred, however, and *O3* was the only one of *K2*'s sons to remain in Kitagata-ami, where he fished with his brother-in-law. My informant regards him as the only one who behaved correctly in the dispute. Interestingly, this marriage and his son's marriage to *T2*'s daughter are the only ones between the two segments of the lineage, whereas cases of *dōzoku* endogamy were frequent. When Norbeck (1954:52) says that adopted husbands usually are not related, or only distantly related to their adopted parents, this is certainly not the case in Shingū where nephews frequently were adopted to marry the daughters. The marriages and adoptions that took place within each segment strengthened their cohesion, but weakened the authority of their respective main households which did not manage to distinguish themselves socially from the households of their segments. Or to quote Nakane (1967:93):

– if a marriage takes place between members of a main and a branch household, it implies the degradation of the status of the main household to that of the branch household. In fact among the minor *dōzoku,* whose main households have not attained any marked superior status within the households of the *dōzoku,* such marriages occur. The avoidance of marriage between main and branch household members shows the degree of power and prestige of the main household as well as the *dōzoku* as a whole. Such a powerful main household tends to have marriage relations with an equally influential main household of another *dōzoku.*

If this holds true, the strongest of the Shingū *dōzoku* was incredibly weak. I believe, however, that these marriages have strengthened the *dōzoku* as a group, and that this is one reason why the Kitagata-ami and Kanmata-ami have survived better than other groups.

Share-holding Net-groups

As the authority of the Matsuo main household grew weaker, the vertical structure that might have existed in Kitagata-ami disappeared and was replaced by a more egalitarian organization. It came to resemble two other big net-groups that emerged from the ruins of Toi-ami and Yorosu-ami. Ichigō-ami (lit.:No.1-net) was first established followed by Nigō-ami (No.2-net).

Toi-ami and Yorosu-ami were probably dissolved in the early thirties, which means that the purse seines were operated by the new groups for a few years. Their main technology was, however, large beach nets. The fishermen were still not prepared to be completely financially independent. About half the shares were owned by farmers who were compensated in fish, but these withdrew after only a few years owing to the lack of profit. For the first time the fishermen owned the important sardine nets without interference from outside.

The share-holding net-groups may be of recent date in Shingū, but it is the only kind of net-group found in the community today. The participants have one share each, and if anybody drops out, he is given no compensation; his capital will fall to the net-group. As shares are not allowed for sale, nobody can acquire more shares than the other, and an economic dominance of a few men is avoided. The same was true when nets were put together in long chains; each shareholder should contribute the same number of nets as anybody else, though a few concessions might have been given to poor households. The egalitarian principles are guarded jealously in Shingū, and this is mirrored in the leadership.

Both Ichigō-ami and Nigō-ami were founded on the initiative of a few outstanding fishermen who had a lot of influence not only among the fishermen, but in Shingū generally. In the initial process they behaved as entrepreneurs, and they were as a matter of course elected as the leaders of the new net-groups. Their authority was unchallenged, but they were unable to transfer their influence to their sons. The leaders were elected according to ability, not descent, and the last step in the democratization was taken when the leader-position began to rotate among the shareholders.

There were from the beginning two kinds of leaders or persons in charge. Two were responsible for the whole operation (*ōyakuin*) while minor leaders (*yakuin*) supervised parts of the operations. The net-groups owned a couple of boats each, and these were used to carry the nets and were therefore bigger than the individually-owned boats. Compensation in economic terms was given to the leaders: the *ōyakuin* received 1.5 shares and the *yakuin* about 1.25 shares. The boat-owners got a third of a share each for their boats' participation.

Though the nets were cooperately owned, it does not mean that everybody had shares. Also the net-groups employed fishermen who did not belong to any group, mostly for financial reasons, but the number of uncommitted fishermen has been decreasing. Employed fishermen received reduced shares as did the women hauling in the nets. In 1975 Ichigō-ami and Nigō-ami still distinguished between shareholders and the others. The former received 1.2 shares, the latter one share each. Kitagata-ami, however, divided equally as the shareholders, in their opinion, were fully compensated through the payments for boats and nets. As there was only one shareholder in each household, additional household members were regarded as employed in this connection, a fact many found unreasonable.

Ichigō-ami, Nigō-ami, Kitagata-ami and Sankichi-ami all operated sardine nets in addition to several other smaller nets. Kitagata-ami operated sea bream nets as did several other groups specially aimed at this fishery. They were of two types; the sea bream beach net of which there were five in 1950, and the sea bream offshore net of which there were three. All were organized on the same principles as the sardine nets outlined in this section, and all are now dissolved.

When the sardine ceased to be hunted by the Shingū fishermen, Ichigō-ami, Nigō-ami and Kitagata-ami (now merged with Sankichi-ami) continued the operations during the yellowtail fisheries. The yellowtail nets are operated by far less people than the sardine nets, but by 1960 the number of fishermen had declined to a level making it possible for everybody to participate.[5] A fourth group (YA-ami in Fig.3) operated yellowtail nets. This group comprised two households of the Matsuo *dōzoku*, but both have been uncommitted since the dissolution of Toi-ami.

Net-groups of this latter kind are probably old in Shingū. There are many cases where two boats have cooperated on the fishing grounds; the two-boats *gochiami* have already been mentioned. As these groups are small, no formal leadership is needed, and they were either based on near kin (agnatic or *shinrui*) or close friendship. Common to most of them is the instability in their structure. They are dependent on an existing relationship between the two skippers. A son seldom has the same friendship with his father's friends; they tended to split when one of the skippers retired. We can find the exception to this rule only when the relationship is renewed in each generation, i.e. when existing affinal ties are reinforced, as was the case in Kanmata-ami and in a few others.

Between 1971 and 1975 one net-group comprising all but a few fishermen operated big trap-nets (*teichiami*). The initiative was taken when four fishermen

Ill.33 A party at the harbour. Such parties always follow occasions of cooperative work.

wanted to buy a net together for a trial. As the trap-net is stationary, agreement from the fishing cooperative had to be obtained and a site had to be allocated for the purpose. To avoid conflicts which were likely to arise as this net hampers other technologies in the area, everybody was to have the same opportunity to participate, and on the initiative of the manager of the cooperative a shareholding net-group was established open to everybody. Most people did join, though women were excluded, and loans were obtained through the cooperative to buy four nets. The group was divided into three sub-groups, and the composition of each follows the rules outlined above, discussing the *taichigyo* fishery. Everybody is entitled to the same share, no compensation is given to the leaders and there are no employees. The net-group owned two unpowered boats bought from Nigō-ami, which both counted as a half man-day, as did the motor-boats. The usage of these latter rotated among the twelve *gochiami* boats. The owners complained that a half man-day was not sufficient to cover the expenses when the catches were low, and the boat-owners were the strongest supporters of the dissolution of the group.

Composition of Net-groups

A fisherman has usually been a member of several management units at the same time. Sometimes he fished within a crew on one boat, sometimes together with a second boat, sometimes with his sea bream net-group, and during the autumn with his sardine net-group. Conflicts between the interests of these groups were bound to develop. Generally the biggest net-group had the highest priority. If it was decided that Ichigō-ami should try the sardine net, it was a matter of course that all the members should follow this decision loyally. The decision was taken either by the leaders or in a meeting, but when the decision had to be taken quickly, as during the sardine and yellowtail fisheries, it befell the leaders to decide.[6] This was probably the most important right of the leaders. The possible conflict was also minimized if the memberships of the groups did not overlap, and this was usually avoided. Thus all the members of Hangorō-ami were members of Toi-ami, as were the Kitagata-ami members; all the members of Kanmata-ami were members of Nigō-ami and so forth. This holds true usually also regarding the two-boats *gochiami.* The only exception seems to have been the sea bream net-groups where the memberships were mixed. One of the sea bream offshore net-groups did not comprise only all the members of Nigō-ami, but also some from Ichigō-ami. In the smaller sea bream net-groups, the mixing of Ichigō-ami and Nigō-ami became more in evidence, but members of Kitagata-ami as a rule never participated in nets operated by other than Kitagata-ami, nor did Kanmata-ami split.

The criteria for membership in Toi-ami and Yorosu-ami are not known, but the membership here has been decisive in the subsequent development of net-groups in Shingū. The members of Ichigō-ami were mostly recruited from Yorosu-ami, while Toi-ami gave "birth" to Nigō-ami. If, as many stated, Ichigō-ami was founded by fishermen not having individual nets while Nigō-ami members had such, this would indicate this distinction also between Toi-ami and Yorosu-ami. I will later discuss the fishing-rights in more detail, but at the moment it will suffice to mention that special rights were required to use nets. To fish by hook-and-line it was sufficient to be classified as fisherman (an inherited status during the Tokugawa period) of the community.

There could be three sources for this dichotomy among the fishermen. Firstly, as one informant stressed, later settlers were not given the same rights as the original

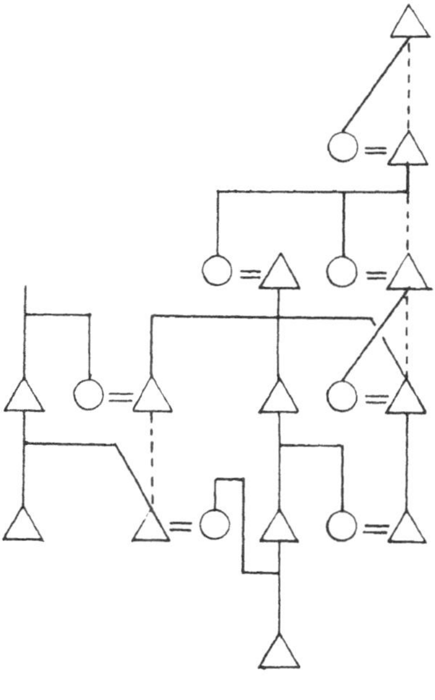

Fig.4 A section of Ichigō-ami.

ones. He also stated that the newcomers mostly settled in the eastern half of Honmachi where Yorosu-ami had a little dominance. The house of Inoue Sankichi was, however, one of the original settlers and is located in the western half, but it participated in Yorosu-ami, as did Yayama. Secondly, the fishing-right (i.e. the rights to fish with certain nets) might have been inheritable, and theoretically branch households could thus be excluded from these rights. This could lead to a certain dependency of the branch household to the main, but it could also lead to branch households fishing with hook-and-line. This would eventually lead to a pattern with the main households in one net-group and the branch households in another, but I have so far found no evidence that this has been the case, at least not since the Meiji Restoration. Thirdly, Toi-ami was based on Toyomura, who was the most influential house in Shingū at the end of the last century, while Yorosu-ami was owned by Sakai in Minato, and he did not have any political influence in Shingū-ura. Toyomura could perhaps have been in a position to influence the allocation of fishing rights. Unfortunately no documents throwing light on this problem have been found.

The membership in the net-groups were, as far back as we have detailed information, decided first of all by descent. There are very few cases where brothers participated in different net-groups, and the sons followed their fathers. Furthermore, the net-groups tended not to intermarry. The cases involving Kitagata-ami and Kanmata-ami are extremes; there are few cases where a Matsuo has married anybody from Ichigō-ami or Nigō-ami. Intermarriages between Ichigō-ami and Nigō-ami were more frequent, but were far outnumbered by marriages within the net-groups. The endogamic marriages within a net-group were, on the other hand, not arranged at random; there are tendencies for a few households to intermarry in several succeeding generations. Fig. 4 shows an example from Ichigō-ami. In this case the resulting affinal ties were not utilized to form smaller net-groups or crew on the boat, but this was often the case. The classical example is Kanmata-ami (Fig. 3), and both Sankichi-ami and Hangorō-ami utilized the *shinrui* relations. There are thus two relations that are utilized when a net-group is formed; the agnatic kin sometimes limited to Father-Son or Brother-Brother relations and the *shinrui* kin, and several such "segments" join in the bigger net-groups. Where the agnatic kinship group is supplemented and strengthened by repeated affinal ties as in Kanmata-ami, a strong and permanent net-group, though small in size, can develop.

At present the net-groups play a negligible role in Shingū. Kitagata-ami and Ichigō-ami were dissolved in 1977. The four yellowtail nets owned by Kitagata-ami were divided among the seven remaining shareholders, while Ichigō-ami's property was divided in two. Nigō-ami still existed in 1978, but it has not been activated for some years.

Whether the declined importance of the net-groups is the reason for the new marriage pattern is doubtful. The same tendency is found everywhere in Japan: the fishermen find their wives outside the community, and there are only a few cases of village endogamy among the younger population. But there can be no doubt that strategical considerations were made when selecting spouses during the days when the net-groups were of great importance. Ichigō-ami and Nigō-ami have always been more "open" in their structures than has been Kitagata-ami, i.e. many of the spouses and adopted sons are from outside the community. On the other hand, these net-groups have been less viable than Kitagata-ami, which was the smallest of the three even before the fission. In 1975 Ichigō-ami had only three active households as members, and they operated two boats (*A10* and *A11* in Fig. 1). Nigō-ami had thirteen member households of which five were from the Kanmata sub-group. (The Kanmata group operated boats *A2, A9* and *B4*, while *A4, B2* and *C2* were operated by other Nigō-ami members.) Kitagata-ami had eight member households operating seven boats (*A5, A6, A7, A8, A13, B1* and *B5*), and YA-ami had two member households with two boats (*A3* and *A12*) and was helped by one member of Nigō-ami, and two households were uncommitted (boats *B3* and *C1*).

Leadership of Net-groups

I indicated at the beginning of this chapter that the status-sets found on the individual boats sometimes could be transferred to the bigger management units, while sometimes new status-sets had to be created. It is important to discuss this point a little further.

Two kinds of cooperation can be distinguished in terms of status-sets. The first are net groups and other forms of cooperation (like the joint marketing cooperative) where the ordinary boats and their crew participate. The nets are either owned individually as during the *taichigyo* fishery or cooperatively as during the two-boats *gochiami,* the *nuikiriami,* and the sea bream fisheries. The status-sets that already exist on the boats are transferred to the new management units, though, as I have already discussed regarding the *taichigyo* fishery, a new status as leader or coordinator was introduced. The second kind of cooperation are the net-groups where the groups do not only own the nets, but also some of the major boats that participate. This was the case particularly during the sardine and trap-net fisheries. These groups owned a couple of net-boats (*amibune*) which were larger than the individually-owned boats as they were meant to carry big nets and many people. Moreover, during the sardine fisheries many of the fishermen participated only on land, pulling the nets. Though the statuses as skipper, core crew member and extra crew member could be found also in this situation, the persons having these statuses might be different from those of the individual boats. Not all the boat-owners could be skippers in the cooperate endeavour. It will therefore be necessary to examine the contents of these statuses in the new setting.

To start with the trap-net fishery, which is by far the simplest and most recent in time: All the participants had one share each, and there were no employees. Moreover, there was no formal leadership when they went to empty the nets. Every night when the nets were emptied, there were two teams; each team emptied two nets. There were two net-boats and their crews were arbitrarily composed. There were also two motor-boats each operated by their ordinary skippers on a rotational basis among the *gochiami* boats. These skippers did not direct the operation on the net-boats, and there was no distinction between statuses during the fishing activities. Everybody knew the operations, but there was a tendency for the elder to give orders if necessary, and an elder usually steered the net-boat to and from the fishing grounds. Nonetheless, there existed a status as leader/responsible (*yakuin*), and two held this position at any time. It was the responsibility of the leaders to decide when the nets should be taken up to be cleaned and

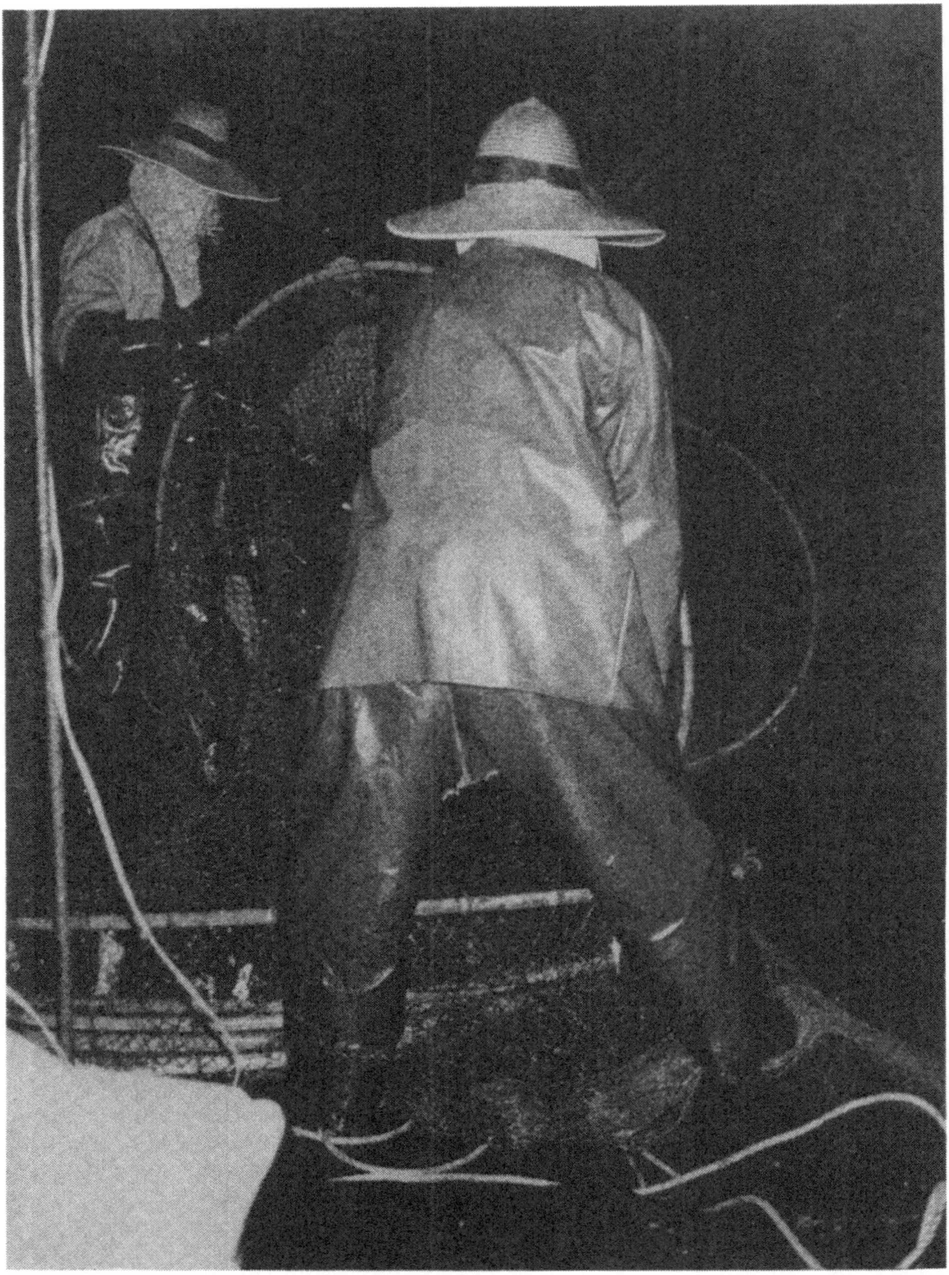

Ill.34 The large trap-nets being emptied.

dyed etc. For this they were given no compensation. The required resource for the leadership was in theory only membership, but again the elected leaders tended to be skippers.

During the sardine fisheries which were of much greater importance, the status-set was more complicated. First, there was a distinction between the shareholders and the employees. About thirty per cent of the catch was distributed among the shareholders for the nets and boats owned by the group. As only one member of a household could be shareholder, the size of the household's contribution counted nothing in this connection. Additional household members were regarded as employees. So were the women and people from outside, either other fishermen or farmers. This brings us to a further distinction in statuses: by "core person" I will understand persons who are professional fishermen belonging to a core crew during the individual fishing. By "extra person" I will understand persons who are employed specifically for a certain job; the women and farmers were included in this category. The extra persons were normally used to pull the nets, as were old core persons. The younger core persons worked on the boats, which was the most physically demanding task.

As we have seen there were two levels of leaders at least in Nigō-ami; two *ōyakuin* and eight *yakuin*. Only shareholders could hold these positions. The leaders did not need to be skippers, though this was mostly the case. As the importance of the net-groups declined, so did the leadership, and the leader's job became mostly an administrative one. The net-groups were reorganized and there was in 1975 only one leader for each of the groups, and this position rotated among the shareholders. No compensation was given, nor was there any prestige involved since they are no longer elected.

But who then, one might ask, took the initiative regarding group fishing? Everybody could take the initiative, and the task was made simple as the yellowtail fishery brings tremendous profits if a shoal is taken. No conflict within the group did therefore develop. If there was any conflict, it developed between the net-groups, and such conflicts have been frequent. In the heyday of the sardine netting battles were fought and nets destroyed. The fishing territory was often divided among the fishermen, but such division did not make sense regarding pelagic fish, and it was important to have many young and strong fishermen who could surround a shoal swiftly and keep others at a distance. As one informant recalled: "The net-boats were crewed by young and strong fishermen, and they were often hot-tempered. The competing net-groups fought at sea, but the elder fishermen settled the quarrels afterwards, and we became good friends on land." Not everybody became good friends, however, and neighbours who happened to belong to different groups did not talk to each other for days. The enmity was latent, and erupted again during festivals, particularly during *Ryūgū* festival when each net-group had its own drinking parties. In this situation it became among the leader's main responsibilities to settle such disputes, but the task was made easier with the intervention of the fishing cooperative.

The distinction between core and extra persons was probably also enforced when the sardine nets were owned by an *amimoto.* Some fishermen had shares in Toi-ami and Yorosu-ami, at least during the final days of these net-groups, and thus had a certain influence which they could use to become the leaders of the operations. The shares were inheritable (they still are) and the position as leaders in these groups also became inheritable. As economic privileges were given both to the owners and to the operational leaders, a certain stability in the economy of these households could be maintained. Whether this gave basis for small *dōzoku* to develop around such leaders is not known, but the possibilities were present.

The owners were, however, mostly found among the merchants. We have seen that the merchants provided capital for the nets, and got a substantial return of the catch. But the transaction between the two parties was more complex than that.

The status as *amimoto* involved a set of rights and obligations towards the fishermen working in their net-groups. The Tokugawan authorities tried to limit the arbitrary dispositions of the *amimoto* and protect the fishermen. A decree issued in Fukuoka in 1772 gave the fishermen the right to use the net even if the *amimoto* ran into economic difficulties, as was often the case. The *amimoto* of the net-groups changed hands from time to time, and rules were laid down to regulate such transactions. The fishermen had the right to use the net even if the *amimoto* had to sell or pawn it; in the latter case the pawnshop was given the share of the *amimoto* (Fukuoka-ken 1963:668). The membership was also regulated by rules of inheritance and descent. In other words: the fishermen were to be secured a place to work, and the authorities did not want the *amimoto* to freely manipulate this.

But the *amimoto* had rights in the form of economic gains. He got a substantial share of the catch, and this was paid partly in kind, i.e. in sardine and other easily traded sea products, thus improving the base of the *amimoto* both in the fields of agriculture (sardine as fertilizer) and as merchant. When cheaper fertilizers were introduced and their trading business declined, they also lost the interest in the net-groups, and this was obviously one of the reasons for the dissolution of Yorosu-ami.

The *amimoto* also had rights in the fishermen, particularly in the operational leaders (*sendō*) who formed the core of the *amimoto.* In the Kujūkurihama District the fishermen entered into contracts with the *amimoto* for a season (the *shoku seido* system), and "once a fisherman entered into the *shoku,* he had to work every day whether or not there was fishing" (Kujukuri Research Committee 1958:103). Moreover, the fisherman was not allowed to leave his master "unless he ran away by night" (ibid.) Also in Shingū the *amimoto* could call on their service almost at any time, especially during the busy seasons like the winter when they often brewed *sake.* On the other hand, in time of need which seems to have occurred almost anually during the winter season, the *amimoto* extended loans to the fishermen. Whether these loans were repaid in full is doubtful. Some of the fish dealers allegedly ran into trouble because the fishermen did not repay their debts. One fisherman told me that he found a lot of receipts from the time of his grandfather stating that fishermen working for him borrowed money for New Year and the Buddhist memorial festival *Bon.* Since they still had the receipts, my informant concluded that the loans were not returned.[7]

The most important resource the *amimoto* needed in order to play his role was, of course, the capital. His position could, nevertheless, be strengthened *vis-à-vis* the fishermen if he also possessed other resources, which he often did. He could control the fishing rights of the community. Each village had exclusive fishing territories, and if an *amimoto* was also the village headman and controlled the village as Ōgawa did in Waki-ura, he could also extend this control to the fishing grounds. This could be a dangerous weapon in the hands of an arbitrary headman. The fact that the *amimoto*, through their ownership of the equipment, also paid the village taxes, further strengthened their power.

The division of exclusive fishing territories, or economic zones, was introduced several hundred years ago to limit conflicts between fishing villages, and probably to estimate the tax-revenues of the fishing villages. Nevertheless, conflicts between fishing villages have been frequent, and a large part of the Akaishi Documents deals with such conflicts over fishing grounds, particularly in Hakata Bay. It was recognized that a sardine net in one village affected the neighbouring villages, and the district headman (or sometimes the magistrate) had to issue licences for such fishing. A merchant was in a better position to influence these agencies than were the fishermen. To understand this it is necessary to consider the administration during the Tokugawa period.

From the villagers the applications went first to the village headman and his assistants, and then to the district headman. This was the case both regarding fishing rights and the applications to borrow money from the *tame-gin* funds. From the district headman the issue went, if necessary, to the military administration represented by the magistrate for the fishing villages (*ura-bugyō*). It was stressed that an issue should never bypass any of these agencies. Herein lies one of the bases for the influence and power of the *amimoto*. As a merchant and *amimoto* he was usually one of the leaders of the village. His possibilities to influence the district headman were better than the fisherman's, and complaints and undesired applications could be stopped at an early stage. Furthermore, the *amimoto* had as a village or district headman judicial authority and could settle disputes over fishing grounds. The important *amimoto* in Shingū-ura all held in periods the position as village headman, except Sakai. Kaneuchi was even the district headman for almost a century, thus increasing the possibilities of controlling the fishing business further.

It will be clear that a successful *amimoto* could at least in theory emerge as a monopolistic power in the fishing villages, and Ōgawa seems to have done so. When this does not seem to have been the case to the same degree in Shingū, comments are needed. The ups and downs of the merchants have already been mentioned. The instability of the *amimoto* is another point to keep in mind. Nakane writes (1967:122):

> Influential wealthy households called *amimoto* (capitalist, and the boss of a fishing team) had to risk their capital, which might be lost in the sea, and also feed in bad seasons a dozen smaller households who formed the fishing team. The ups and downs of these *amimoto* households are a well known story.

Though this might be particularly true for an *active amimoto* who had no other resources, it certainly did not stabilize the economic base of the merchants. A final point should not be overlooked, and this is the internal structure in Shingū-ura. There are five arguments against any strong and dominating *amimoto* in Shingū, at least after 1800:

1. There existed active fishermen who were *amimoto* at least to the sea bream nets.

2. Many fishermen had individually owned boats.

3. The position as village headman seems to have rotated among some of the households of the merchant class, thus undermining the basis of a monopolistic *amimoto.*

4. The fishermen were represented in the shrine committee on an equal footing with the merchants as each street elected or appointed two representatives to this committee.

5. The opportunities of the fishermen to play an active part in the village affairs in the Tokugawa period are not known, but as early as the Meiji era some of them played prominent roles.

The fishermen, at least some of them, participated in the communal affairs, and no house had any monopoly of any of the offices, with the exception of Kaneuchi as district headman up to about 1820. It may seem that Shingu-urā was what Izumi and Gamō (1952, cited in Johnson 1967:156) call an egalitarian village, and this brings us to a discussion of vertical contra horizontal structures in the fisheries of Shingū.

Nakane puts forward in her *Japanese Society* that the groups can be based on two opposite criteria; frame and attribute. Frame includes locations like a village and a company, while class and caste are based on attribute. A group based on frame can include individuals with a great variety of attributes, while a group based on attribute can be located in many localities. Thus a boat is a frame, and the crewmen possess several attributes (or statuses as I have called them). Nakane continues that corporate groups based on frame "is the basic principle on which Japanese society is built" (1973:7). Nakane further postulates that a group based on frame tends to be vertical in nature and those based on attribute horizontal. Hence her conclusion that the basic characteristic of the Japanese social structure is the vertical group.

The vertical group is characterized by a set of dyads linking the leader and his subordinates. It becomes important to knit the members in vertical relations, and to secure that one and only one of its members always holds the position as leader. This is achieved through a rigid ranking system where the seniority principle is the most important, and where personal abilities count less. The members are united in a network of super-subordinate relations according to seniority and where one's peers are regarded as potential enemies.

The recruitment to the vertical group is, according to Nakane, from the bottom. New members are knit to existing members in direct super-subordinate relations that can develop into *oyabun-kobun* ties. In such a group the change of leader can lead to a critical phase of the whole group. Often, particularly when two potential new leaders have the same seniority, a struggle for power will develop, and their respective subordinates will be mobilized. The outcome can either be factions, which is clearly exemplified in the Japanese political parties, or to a fission exemplified in the Kitagata-Kanmata struggle. We can understand the success of the entrepreneurs in the creation of Kanmata-ami in this light. The only one of seven brothers who did not leave Kitagata-ami was either already a client of his father-in-law who was his father's enemy, or because of his marriage, hoped for a reconciliation between father and daughter, and in order to achieve this he demonstrated his loyalty by remaining inside Kitagata-ami. At any rate, he succeeded in becoming a client of his father-in-law and became a core crew member on his boat.

The fissions that take place on the boats can also be understood in this scheme. As long as the father is active, the crew will remain stable, but the brothers might later compete to take their father's place. This was particularly common, as was shown in Chapter VI, when the brothers possessed the same resources and experience.

The leadership problem can be avoided in a vertical group by institutionalizing the succession. Nakane points out that there is an upward mobility within the group by adding new members at the bottom, but this does not always have to be the case. Groups can be stratified so there is a minimal upward mobility because each level is self-recruiting through, for instance, descent. This is the case in the leadership in many family-based companies, the *iemoto* system in the schools of art, and in some of the religious movements. And this was also the case in the net-groups based on *absentee amimoto,* the structure of which is shown in Fig. 5.

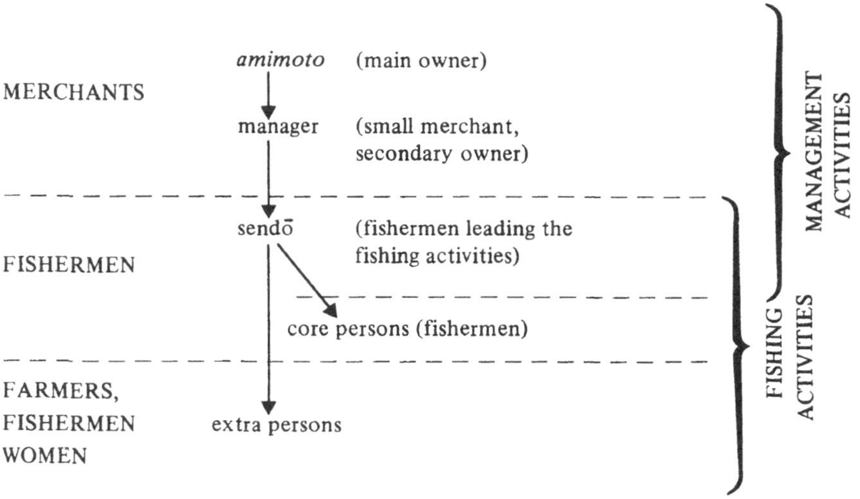

Fig. 5 The Structure of a Net-group based on *Absentee Amimoto.*

The upward mobility within this structure was very limited. The position as *amimoto* was inherited as a part of the household property, but could through sale be transferred to another *amimoto.* The manager was also mostly from the merchant class, but whether he was elected or inherited his position is not known. The status as *sendō,* though held by fishermen, was also allegedly inherited. Ideally, the only mobility that existed was that of succeeding sons who started as core persons and climbed to *sendō.*

The group as well as the leadership is stable and crises are avoided. The same is true in the *dōzoku* as long as there is no challenge to the leader (*honke*), neither politically nor economically. In such circumstances a *dōzoku* based net-group might also be stable. (The problems arise when the main household is weak and quarrels erupt as to who is the actual main house.) The stability is secured by a careful ranking of the group members and where the recruitment to each rank is institutionalized through inheritance. On the other hand, the dyads between the *absentee amimoto* as *oyabun* and the fishermen as *kobun* became perhaps less intimate. An *absentee amimoto* was not required to possess the same charismatic abilities as an *active amimoto*, who could not depend on a well established structure outlined above.

Nakane's theory can explain many facets of the Japanese social structure, but by turning all her attention to the vertical group, she tends to ignore the horizontal ones, though she briefly admits the existence of these. Nonetheless, horizontal groups are important, and they are neither new creations nor unstable.

A horizontal group is characterized, according to Nakane, by strong relations between peers, which will of course influence the position of the leader. The second characteristic of these groups is that they are not necessarily ranked. The problem is therefore how these groups manage to have an efficient leadership and still remain unranked. Nakane postulates that "what is most important is that in Japan a group inevitably and eventually develops the vertical type of organizational structure" (op.cit.:65). But this is, as will now be shown, not always the case.

In a ranked group leaders can easily be appointed as long as there is no competition between several persons of the same rank. An appointment will, if it is done according to ranking, not change the structure of the group. If, on the other hand, the members are equal in standing, it will be difficult to appoint new leaders without generating new statuses expressed in rank. Where equality exists and the members are not willing to alter this, a conflict arises between the desire to have a leader for the operation and who can take the responsibility on the one hand, and the desire not to create differentiations between the members on the other. This dilemma can be solved if rotation of the leadership is institutionalized, without considering the personal abilities of the members. A minimum of responsibility is therefore given to such a leader, and he is stripped of all prestige and economic compensation. The leaders of the net-groups in 1975 did little more than accounting. Where more is required of the leaders (as during the *taichigyo* and pre-war sardine fisheries), personal abilities are considered and there is a secret ballot. To secure a certain rotation also in this case, rules are made to prohibit re-election. Some prestige is involved through the election, though everybody stressed that this was not the case. Nonetheless, some were quick to announce their elections, and they liked to give the impression that the job was demanding and required a person with only the best qualities. It was mostly young leaders who bragged about their elections, which can be regarded as an official recognition of their status as able skippers.

Ranking is hard to avoid in economic terms where the competition between the fishermen is as hard as in Shingū. Some are able to convert their success as fishermen into some influence, but at present such ranking is discouraged in the cooperative fishing like the *taichigyo* fisheries. This is achieved by keeping the catches secret. Some disagreements exist as to the desirability of this secrecy, and some rather successful skippers wanted to abandon this in order to encourage competition between the crews, and thus – so they said – increase the overall catch.

The rotational leadership is perhaps a relatively new phenomenon in Shingū's fisheries, though information on this point before the Taishō era is meagre. It might have developed in Ichigō-ami and Nigō-ami in the 1930s, and can thus not be accredited any post-war popular democratic movement. The inspiration came from within the community, and the rotational system is far from new to the Japanese. It can be seen clearly in the formal administration of communal affairs. The position of the village headman in Shingū during the Tokugawa period has been mentioned. At that time the headmanship was restricted to the upper echelon of the population, while the position as ward headman (*kuchō*) has, since the beginning of this century at least, rotated among the three streets. The position as leader of the neighbourhood groups (*kumichō*) is further rotated among all the households in the group according to the location of the houses, though at least Honmachi-1 rotated this position according to age until recently. Some informants stressed that the rotational responsibility was also valid in the election of temple and shrine representatives. For example, Honmachi West (Honmachi 4-5-6) elects one representative to the shrine committee, and this should rotate among the three neighbourhood groups. This procedure was unknown in Honmachi East, however, and not adhered to in Honmachi West either. Nevertheless, the idea of rotation was a part of their "conscious models", and this principle is indeed well-known in the literature on Japan, and is as "typically" Japanese as the vertical groups.

Two principles are in force in the Shingū net-groups; the vertical and the horizontal. And they do not necessarily exclude each other. The vertical structure in Toi-ami and Yorosu-ami gave way to a more horizontal structure in the new net-groups that developed, and the decisive factor in this development seems to have been the ownership of the capital. On the other hand, both principles can be seen working hand in hand during the *taichigyo* fisheries. Each boat can be regarded as a frame and can (with the modifications made in Chapter VI) be looked upon as a vertical group. Ideally the status as skipper should be unchallenged. On the other hand we have the horizontal principle in the leadership of this marketing cooperative. No ranking was allowed to develop between the boats, i.e. between the skippers.

This point bears out my argument. The skipper should ideally possess resources his crew members do not possess. He controls the capital, and the fishing licences are allocated him by the fishing cooperative. This tends to create vertical structures. When it comes to the leadership of the marketing cooperative, the egalitarian principle *among the skippers* is stressed since the skippers all control the same resources. Care is taken that no resource is monopolized and thus generates differences and vertical structures among the skippers. The fact that all the skippers who were eligible to the leadership position of the marketing cooperative received votes, illustrates this. The desire to rotate the leadership is stronger than the desire to vote the best man into the position.

A sub-group based on the attribute "skipper" was formed, and this worked horizontally. A sub-group based on the attribute "crew member" has not been created, and it is undesired from the point of view of the skippers. Such a group might seriously influence the skipper's position on the boat and undermine the loyalty between him and his crew. It is also undesired from the point of view of the crew member. His competitor is not his skipper but the other boats. The stress on competition can also be seen in this light; it strengthened the in-group feeling of the crews at the same time as it clearly distinguished between "us" and "them".

I have in this and the preceding chapter explored the different management units, their structure and social organization found in Shingū during the last hundred years or so. The next step is to relate these management units more firmly to the changes that have occurred regarding the control of capital, fishing rights, markets and labour, and this will be the theme of the following two chapters.

CHAPTER VIII

ORGANIZATIONAL CHANGES IN THE FISHERIES

The control of the basic resources – capital, fishing rights, market and labour – has undergone important changes since the Meiji Restoration. The control of the labour force has already been discussed in parts (Chapter VI), and I will again return to this in the next chapter. In this chapter I will mainly discuss the three other resources and how the changes in the control of these have influenced the structure of the management units.

The Situation at the end of the Tokugawa Period

I will start with a recapitulation of the situation at the end of the Tokugawa period. Many of the fishermen owned small boats and some equipment, usually hook-and-line but also smaller nets. Among the important species caught by these technologies were the sea bream, mackerel, horse mackerel, squid, *konoshiro,* yellowtail and pig fish. Many of these boats were operated by a group of agnates, i.e. father with sons. The mean household size was on the other hand as low as four or five towards the end of the period, which indicates that many households were unable to man their boats without hiring fishermen outside the households. Cooperation between households might also have occurred. Fishermen with no capital to invest had thus an opportunity to be employed as core crew members, while farmers and poor pedlars might have found seasonal employment as extra crew members.

The fishermen also participated in large net-groups which were often centred on *absentee amimoto* who were dominant merchants. It was particularly the sardine fisheries that required this kind of organization as this species was caught by large nets, but other species like sea bream and grey mullet were also caught by big nets. The merchants furnished the fishermen with funds for nets and net-boats, and in return they did not only receive a share of the catch, but also established ties of loyalty to many of the fishermen, ties they could utilize for a multitude of purposes.

The control of the fishing territories was an important issue throughout the Tokugawa period. The fishing rights became formalized because of the increasing commercialization (Bernier 1970:178) and for taxation conveniences. The fishing villages had to pay taxes both in money, kind and corvee labour, and only villages which sent *kako* were given territories and rights (Nakamura 1961). Only fishermen of the community were allowed to fish inside the territory, but also they met restrictions. The number of fishing rights were limited, and the village leaders (usually the merchants) had a decisive influence in the allocation of these rights. Moreover, the fishing rights became increasingly private property and could be sold, pawned and leased (Seiko 1966:98-99).

Little is known about the marketing of fish in Shingū-ura, but some tentative suggestions can be made. The fish fell mostly in two categories; fresh and processed. The fresh fish had to be brought to the consumers fast, but as the transport system was poorly developed, each dealer could only carry small quantities of fish. This might explain the high number of non-fishing households in Shingū-ura. A high

Ill.35 Straw protected the old nets before the introduction of nylon nets.

number of pedlars brought the fresh fish to the consumers. The circulation of processed sea products was probably more restricted. The situation regarding the sardines has already been mentioned; the income from the sale of this was supposed to be used to pay the village tax. Dried abalone (*awabi*) served a similar function (Tsukushi 1974:113). The Japanese fish paste *kamaboko* was also manufactured in Shingū and was a niche for some merchants. The fish then gave rise to small-scale industry and trade within the village. Fish was transferred from the fishermen to the merchants, but it is not known exactly how this was done. Two approaches can be suggested:

First, the merchants and fishermen established fixed trading partners, and the former went into a partnership with the latter. Already in the beginning of the Tokugawa period Ōsaka sardine wholesalers advanced "silver to the fisheries in various parts of the country for them to purchase materials for fish nets, and established an interest through *kabu* (membership privileges) in new fisheries and new fish net producers. The sardines which were caught were salted and sent to the *tonya*" (Sheldon 1958:82). The "putting-out" system worked in a similar fashion, but the merchants advanced raw material and equipment instead of money.[1] Instrumental to such partnerships were loans advanced to the fishermen, thus putting them into indebtedness, and the merchants obtained a steady and cheap resource for other activities like agriculture and trade.

Second, market places developed and the transactions were at least partly governed by supply and demand. This was the prevalent form in the Meiji and Taishō eras when net-groups based on *absentee amimoto* still existed. The market place was operated by a wholesaler (*tonya*) who arranged auctions. The fish was then sold to fish-dealers (*nakagai*) who brought the fish into the countryside, to shops, to other markets with better prices, or processed the fish themselves.

Perhaps both forms of exchange have existed side by side as they do not necessarily exclude each other. One species could have been sold by auction, while another was channeled to fixed trading partners. Or a part of the fish was given as rent to the *amimoto* while the fishermen's shares were sold at auctions and so forth. It is reasonable to assume that the auctions became increasingly important after the Meiji Restoration when the taxes, which were formerly paid partly in kind and corvee labour, now had to paid in money only.

Such then was probably the situation at the end of the Tokugawa period. The merchants had some possibilities to control the resources; the fishing rights through the office as village headman or his assistants, the nets and the markets through their capital. But fundamental changes occurred with the turn of the century.

Fishing Rights

A new fishing law which was introduced in 1901 transferred the fishing territories from being the corporate property of the fishing villages to fishing associations (*gyogyō kumiai*) that were established all over Japan. By one stroke the position of the *absentee amimoto* was theoretically challenged. Some of the fishermen acquired more influence both on the territory as such and on the licences to be allocated, and the influential village leaders were not able to manipulate these so easily as before. Nevertheless, some fishermen could, through their influence in the association, manage to manipulate these licences as the fishing associations were not necessarily open to all the fishermen. In Shingū an estimated forty per cent were excluded either because they could not pay the initial fee, or because they were new settlers. Moreover, the licences were allocated often to named persons, and they could be inherited, lent and even sold (Shibusawa 1958:364).

In 1935 there were three kinds of fishing licences in Shingū: (1) *Senyōgyogyōken* which allowed the fishermen to conduct general fishing within Shingū's territory. (2) *Teichi gyogyōken* which allocated certain areas to fixed equipment like trap-nets etc. This kind of licence could be leased. 3) *Tokubetsu gyogyōken* which allowed certain fishermen to conduct "special fishing" such as sardine and sea bream netting. The leaders of the big net-groups needed this licence, but they could lease these to other members of the net-groups.

The laws were again changed in 1949, and the fishing territories became solely controlled by the cooperative. The Shingū fisherman can only use the licences allocated him by the cooperative, and they have to be returned if not used.

The village merchants not only lost the influence in allocating the fishing licences, but also the possibilities of mediating in conflicts between fishing villages over fishing rights. The fishing territories have remained exclusive, but illegal trespassing occurs quite frequently. While the village headman or more frequently the district headman settled such disputes in the past, this authority now usually belongs to the fishing cooperatives that have institutionalized a way to settle these disputes.[2] Other sources of judicial authority were after the Meiji Restoration removed from the headmen and given to a state police and the courts, thus depriving the headmen of their possibilities for legal sanctions in the villages.

The territorial rights today are classified as fixed fishing rights, sectional fishing rights and common fishing rights. The sectional rights are the rights to undertake fish breeding and sea-weed cultivation on sections of the territory decided by the cooperative. The common fishing rights give the fishermen the opportunity to catch and collect all kinds of aquatic products provided he holds a fishing licence where such is required.[3] Several of the technologies need special licences, however, and this is particularly so regarding newly developed technologies.[4] Shingū controls at present twelve licences for *gochiami* fishing, and all these licences are used. Shingū further has eleven licences for yellowtail netting, three licences for mullet netting, fifteen licences for ivory shell trapping and so on. If a new boat wants to

Tokugawa period	Present situation
fisherman	fisherman
↓	↓
village headman (= *amimoto*)	Shingū Cooperative
↓	↓
district headman (= *amimoto*)	Fukuoka Fishing Dept.
↓	↓
(magistrate)	Ministry of Agriculture and Forestry (for offshore fishing)

Fig. 6 The procedure taken when applying for Licences.

participate, an additional licence has to be obtained. This can theoretically be done in two ways. It can be leased from another cooperative within Chikuzen area of Fukuoka Prefecture if this cooperative does not utilize all its licences. Or, if this possibility does not exist, Shingū cooperative can apply to Fukuoka Fishing Department for another licence, or to the Ministry of Agriculture and Forestry in case of offshore fishing. It is, however, extremely difficult to be allocated new licences as the maritime resources are already heavily exploited.

If we compare the procedure taken when applying for a fishing licence during the Tokugawa period with the one taken today (Fig. 6), it will be clear that the *amimoto* could manipulate the decisions particularly if they also held positions as village or district headmen. Today they have no influence as they have no access to the cooperative. That the active fishermen within the cooperative can influence the decisions is another matter which will be discussed later.

Capital

The fishermen have steadily improved their influence on the capital and investment in the fisheries since the turn of the century. The success in establishing their own independent net-groups on the ruins of the old has already been explored. These nets were financed by pooling their assets, all contributing equally. Those who were too poor to contribute were, depending on the circumstances, either accepted as members (but with reduced influence) or they were employed. The fishing co-operative started only later to extend credits, and in 1971 all the capital needed for the large trap-nets was borrowed through the cooperative. This enabled all those interested, regardless of economic standing at the time, to participate.

The cooperative does not only extend credit. Sometimes undertakings that are hoped will improve the living standard of the fishermen receive financial assistance in the form of subsidies. When the equipment was bought for *wakame* cultivation, the fishermen received half the cost. The offer was open a couple of years to every-body who wanted to try.

The cooperative is encouraging all kinds of cooperation and new invention, but it is more restrictive regarding privately owned boats and equipment. Usually these were financed by the fishermen themselves, though small loans were given from the *tame-gin* fund during the Tokugawa period. They borrowed from relatives and from credit associations called *gyosenkō* or *funakō* (lit. boat credit association).[5] It seems that these *gyosenkō* first developed in Shingū during the Taishō era, and there were two such associations, each centred around a boat builder.[6] Everybody could in theory become a member, but as they had to pay a monthly fee of a couple of yen, some could not afford it. To avoid deficits, people's ability to pay regularly was taken into consideration, and this made it more difficult for persons who owned no boat to become members.

The operation of the associations was rather complicated. Sticks were put in a cup, and each member with drawing rights took one stick. (Those who had already won could not participate any longer, though they had to pay the monthly fee.) A number was labelled to each stick, and the one who got the winning number announced on the wall or otherwise, received the lot *honkuji.* This lot was a long way to finance a boat that cost about 120 yen in 1925, but parts of it were used to pay for tea at the meeting and for *sode*, that is a little money given to each on either side of the winning number. It was therefore desirable not to win in the beginning because the winner was later excluded from drawing sticks and thus receiving *sode*. The last winner did of course pay no *sode* whatsoever. Some fishermen who needed money badly could, on the the other hand, purchase that month's lot for a fixed fee, usually an additional fifteen yen. It is therefore fair to say that these associations benefited those who were better off financially because they could afford to wait for the winning lot. Or if they won, they could sell it to someone in need. Nonetheless, this institution did open a possibility for upward mobility, as employed fishermen who showed industriousness and were found creditworthy were, theoretically at least, allowed to enter the association. It also admitted a branch household with the main house as guarantor.

The fishing cooperative allegedly encouraged the formation of these associations, but their days were ended after World War II. The direct cause was the introduction of the engine. The boats became too expensive to be financed in this way. Furthermore, with the introduction of these new boats, the local boat builders could not compete with the yards in Nata, Tsuyazaki, Ainoshima and elsewhere. Today the boats are built as far away as in Ōita, Nagasaki and Kumamoto prefectures.

After the abolition of these associations, the cooperative became more active in financing private boats, and the first loans for such boats were given already before the war. The loans granted were dependent on the fishermen's catches, and they had to be guaranteed by the members of the executive committee (*yakuin*). It was common for a fisherman to visit such a member carrying a bottle of *sake* under his arm. Today this guarantee is depersonalized, and the influence of the *yakuin* is reduced.

Several kinds of loans are now available to the fishermen through the cooperative. The cheapest are the Modernization Funds (*Kindaika shikin* and *Kōzō kaizen shinkin*) with interests of 4.5 and 3.5 per cent respectively. The maximum amount in 1978 was five million yen or eighty per cent of the total cost, which is to be paid back in fifteen years. More expensive are the *kentan shikin* loans with an interest of 9.5 per cent, but these loans are sometimes used in emergencies because of the bureaucratic procedures taken when applying for the cheaper loans.

A new *gochiami* boat on the other hand cost almost eight million yen in 1975, and the owner has to finance the rest himself. This is usually done through the saving accounts at the cooperative, and such an account is also usually necessary in order to get the guarantees for the loans. Thus, it is much easier for one who has a boat to finance a new one than for an employee to buy his first.

Again we can see that those who already own boats are favoured, but they all have the same access to capital, the only determining factor is the ability to catch fish.

Market

At the beginning of this century the fish was sold at markets in Shingū under leadership of auctioneers, the *tonya*. The *tonya* got a certain percentage in commission, and this was an incentive to force the prices upwards. A few of the smaller *amimoto* were also *tonya*, but there is no indication that the large *absentee amimoto* Yorosuya and Toyomura were *tonya*, at least not in this century. (The situation might have been different during the Tokugawa period.) It was allegedly expensive to be *tonya*, and they changed frequently. He had to pay the fishermen before he received his money from the fish-dealers (*nakagai*), so a certain capital was necessary (cf. Wigmore 1969:94). He was therefore one of the leading merchants in the community. Moreover, the fishermen often went to borrow money from him when in need; money which was often difficult to get back. Formerly the *tonya* might have been rewarded with the partron status *vis-à-vis* the fishermen, but when the opportunities for other employment improved, the *tonya* lost many of his possibilities for sanction. Moreover, as his own business in Shingū declined or became based on regular employees, the patron status had little relevance even for him.

Obviously a more contractual basis had to be developed if this institution was to survive. However, it did not get time to be reorganized. About 1925 the *tonya* was the brother of one of the fish-dealers. The fishermen accused him of favouring his brother, and for not trying to force the prices up. The outcome was that the fishing cooperative went into the marketing business and established its own market in Shingū. The cooperative could, however, hardly compete with the private market allegedly due to the higher administrative costs. The private market nonetheless closed after an appeal by the cooperative a couple of years later. By this move the last relations (except the work in the *sake* breweries which continued for another ten years or so) with the merchants were curtailed.

The cooperative got a whole new range of activities because its economic base was broadened. Previously it had been merely an association allocating fishing rights. The market in Shingū was on the other hand small, the fish-dealers were not many (twenty-three in 1949) and the supply of fish was unstable and one-sided. The fish has since the 1960s been brought to the much bigger Hakata market which is operated by a private *tonya* taking a commission of five per cent in addition to the six per cent paid to the cooperative. If the fishermen sell their products directly to consumers, as they do with *wakame* and some minor species, they should still pay the six per cent, but it is not always done though such behaviour is sanctioned by gossip and accusations.

Some species are never taken to the Hakata market; the sardine was always sold in Shingū as was the sand lance. The ivory shell is at present sent directly to Tōkyō where the prices are better. (The shell can be kept alive for a week so the quality is not reduced by it being sent to Tōkyō.) The salted *wakame* is either sold directly by the fishermen or by the cooperative to consumers. And today the *taichigyo* is sold directly to the fish breeders. This has been a hard issue during the last couple of years.

Shingū, which was a pioneer in this fishery, built up a range of customers of which the giant Taiyō Gyogyō is the most important. But Fukuoka Fishing Cooperative, which is the central cooperative for the prefecture, decided that all the fry should be marketed by them, and a commission of three per cent should be levied in addition to the ordinary six to the local cooperatives. In return the central cooperative would guarantee the payment of the fish; sometimes the fishermen did not receive their money either because the buyers have meantime gone bankrupt or for other reasons. Shingū was strongly opposed to this arrangement. They had managed to get reliable customers and no guarantee was therefore felt to be needed. The prices Shingū received from Taiyō Gyogyō and others, were much higher than the prices the central cooperative managed to get (90-100 yen a piece against 60-70 yen in 1975), because of the high quality and the long trading relations. Shingū was also against the arrangement for two other reasons: they did not want to pay the extra commission, and it would be harder to manipulate the tax papers. From 1975 Shingū was again allowed to sell directly, but the commission should be paid.

The same trend can be observed in the development within all the three resources discussed: the capital, the fishing rights and the market. The merchants lost their control of these resources, and there are signs that they gave away at least some of these resources voluntarily due to changes in the infra-structure. The fishermen were "left" in poverty, and there was agreement that they grew poorer during the Taishō and early Shōwa eras. The fishing cooperative came to take over the functions of the merchants, and it could be tempting to conclude that the fishermen have, through their membership, gained influence over the decisions concerning these matters. But before we can make such a conclusion, we have to look at the internal structure of Shingū Fishing Cooperative.

The Organization of Shingū Fishing Cooperative

The policy of Shingū Fishing Cooperative (*Shingū gyogyō kyōdō kumiai*) is, within limits set by the authorities through the Aquatic Products Cooperative Law, decided by an elected committee of six members (*yakuin*). This is headed by a chairman (*kumiaichō*). Previously the office as chairman was of great importance, and the competition for this office could be fierce, as it still is in other communities where the cooperative is the basic organization of the whole hamlet. The chairmen in the past were often young and ambitious, and the office was an important asset that could be used in running for other positions, for instance to the village or town councils. The *yakuin* were usually among the leaders of the net-groups. In 1936 the chairman was also one of the leaders of Nigō-ami, both the leaders of Ichigō-ami were also *yakuin,* Inoue Sankichi – the leader of his own sardine net – was *yakuin* as was another influential person in Nigō-ami and one in Kanmata-ami. The only group not represented was Kitagata-ami, and it seems that this group has long been under-represented. The reason for this is unknown, but it might have connections with the fission of the net-group some years before.

Today there is no competition; on the contrary, it is difficult to find someone willing to take the job as chairman. The financial incentive is a meagre 250,000 yen annually (1975), and the position is demanding as there are frequent meetings with every level of authorities. He therefore needs time and additional sources of income, and it suits best a person who has retired or is self-employed and able to adjust his working hours to fit with the duties as chairman. This can hardly be done by an active fisherman who has to adjust his operations to the weather. When the tasks of the cooperative were more limited, the contacts with the other agencies were also more limited and it was easier for an active fisherman to combine his occupation with the office as chairman.

In 1975 the chairman was an active fisherman (as all the chairmen have been except one who left the fishing boat at an early age), but for the first time in

Ill.36 The three employees at work in the Fishing Cooperative.

Ill.37 The new building of the Fishing Cooperative.

history he is not a skipper, nor a boat-owner. He is employed on his nephew's boat which was the least active of the *gochiami* boats, and which was beached during the winter. Moreover, he was not the important money-earner in his household, as both his wife and married son have permanent work outside. The chairman is then elected on the criterion of time to spare, not primarily on personal abilities, and this has implications for his role performance. The younger fishermen accused the chairman of being passive and weak, i.e. he was busy keeping the cooperative running "as smoothly as possible according to old customs", while they wanted him to act more vigorously fighting for and testing out new ideas. He is, according to some, also heavily influenced by the manager of the cooperative, and through this influence the manager has a crucial position in forming the policy of the cooperative.

The daily routines and the policy outlined by the committee are carried out by three full-time employees: the manager, an executive and a secretary. The manager and the secretary are born in fishing households and both have brothers who are active. The areas of responsibility are clear-cut in theory, but it cannot be denied that the manager in particular plays an active role in forming the policy of the cooperative. He can participate in the committee proceedings, and in the absence of the chairman, act as his deputy. The interplay between the manager and the chairman is of course heavily dependent on their personalities, but the manager's position is strengthened by the fact that he is permanently employed while the chairman is elected for a couple of years. This gives him an unquestionable advantage in making contacts beyond the cooperative, and he can utilize these to improve his ability to influence the members. Furthermore, he has access to knowledge which the fishermen only seldom get. This knowledge involves rules regulating the fisheries, but also the knowledge of accounting. This is an important asset, not only to keep the economy of the cooperative in order, but to help the fishermen manipulate their tax-reports. His limitation is first of all that he is employed only as long as he has the fishermen's confidence. The former manager lost this and was forced to quit. His position is thus more fragile than that of the *amimoto* (who also was in charge of the management activities) who could not be removed easily.

There are two kinds of membership within the cooperative, the full member *(seikumiai-in)* and associate members *(junkumiai-in)*. There are forty-six regular members of which only twenty-nine were active in 1978, and two associate members, none of them active. The main difference between the two kinds of members is that the associate member can neither vote nor be a candidate for an office. They have also fewer shares than the regular members. In Shingū they have one share each while the regular members have between 260 and 11 shares at 1000 yen.

All the regular members have theoretically the same rights regardless of the number of shares they possess. In practice the non-active members do not participate in the meetings at the cooperative, and among the active members all but a couple have 260 shares.

Strictly speaking, there is a regulation stating that only active fishermen can be members of the cooperative. This is to exclude non-fishing capitalists to gain a foothold. On the other hand, only members are allowed to fish. Why, then, are there so many non-active members of the cooperative?

Shingū fishing cooperative is, with its forty-six regular members, one of the smaller cooperatives in Japan, and this has both political and economical consequences. It is the policy of the authorities and the central cooperative to encourage amalgamations of smaller cooperatives. It is believed that the economic base, and thus freedom of action, will be improved by this move. The fishermen of Shingū are nevertheless afraid that they will be dominated by the much bigger cooperative in Ainoshima, and be deprived of the resources they control today. Hence they are against the amalgamation; they try by all means to give the impression of an active and big cooperative. Big delegations of five or six are sent to most meetings with the authorities – even to meetings on the national level in Tōkyō. By the same reasoning, people who have stopped fishing are still encouraged to remain members. Economically this is also found advantageous from the cooperative's point of view. If they withdraw from the cooperative, their shares have to be paid, and this is a considerable amount of money. Furthermore, in case of deficits they are requested to contribute to the balance as the active members.

Not all the fishermen are members of the cooperative, however. As a rule there is only one member in any household, and this member is usually the father or elder brother, though all are members of the fire brigade *Kyūsaikai* (if younger than about forty-three years) and can participate in net-groups on equal terms.

Nonetheless, for several young fishermen it feels frustrating not to be able to participate in meetings that concern them, and they are excluded from any elective office in the cooperative. The members of the executive committee tend therefore to be rather old; at present between forty-seven and sixty-two years old. The "young" members are elected separately to represent the young generation; their ages were forth-six and forty-seven in 1978.[7] The problem is accentuated as most of the present *yakuin* do not have sons who are willing to take over the business, and they are consequently accused of not considering the future of the fisheries in Shingū. This issue was particularly in focus when the cooperative ran into economic difficulties. Fishermen who claimed they were willing to invest for the future wanted the deficits to be paid by the members. Others wanted the deficits to be paid by sale of the sand from the beach. The latter course was taken and sand was sold not only to cover the deficits, but in such quantities that a profit could be distributed to the members. Again the conflict arose: should this profit be added to the capital or should it be distributed? The opponents of the sale argued that the latter course would imply an incentive to further sales, and thus further undermine their own resources. It is believed – a belief that might find support in empirical facts – that the exploitation of the sand-reserves leads to worsened spawning grounds for fish (particularly for the important *kisugo*), decline in the number of shells and squids, and increased frequency of broken equipment due to uncovered stones and other obstacles on the bottom. But the opponents lost the struggle as they were under-represented in the decision-making meetings where only the regular members had access.[8]

One further point brings the conflict to the surface. If the deficit should be turned to profit (as was done in the sand case), non-active members will receive more money than the active non-members. This key of distribution is also followed when compensations are paid to the cooperative for inconveniences caused by the authorities. When the navy has military manoeuvres in the area, compensation is given, which so far has been added to the capital just to avoid the conflicts that would erupt if it was distributed as outlined. The same holds true if Shingū should be hit by cases of pollution. Understandably, the active non-member thinks it unfair that a person not affected at all should receive more compensation than himself, whose whole lifelihood is threatened.

Two suggestions were made by the fishermen themselves to solve this latent conflict. The first was to change the status of the non-active regular members to associate members. This would make it possible for them to return to fishing in the future if they so desired. The other approach, which has been adopted by the cooperative in Ainoshima, was to allow several members within the same household. This would also bring in new capital. On the other hand the new members would also be entitled to loans which the manager in Shingū finds problematic.

As the fishing community in Shingū is small, all the active members meet almost daily, and they can discuss issues at stake. In practice, however, the opportunities to influence the decision-making are limited. First, because of the distinction between active members and non-members outlined above, by which the access to influence is institutionalized. Secondly, there is a distinction (not officially) between those who fish with *gochiami* and the others. The latter is a small minority and, moreover, they contribute less to the economy of the cooperative. They feel, particularly the elder ones, that they do not have the right to voice their opinion, and a few of them avoid the cooperative altogether. They did not protest when the majority made an agreement with the cooperative in Nata. The *gochiami* boats from Shingū should be allowed to fish in Nata waters, in return five Nata boats were allowed to keep living sea bream fry in nets in Shingū harbour since the harbour in Nata was too shallow, and Nata gill-netters were allowed to fish in Shingū waters.

This issue was brought to the surface again when the concern Taiyō Gyogyō, which is Shingū's best customer for *taichigyo,* wanted to hire a part of the harbour to breed fish. From the very start the interests of the members were divided. The majority wanted to welcome the company for two reasons: it would guarantee to buy at least one third of the fry, and they felt some obligations towards the best customer. They were afraid of losing this customer if they turned the proposal down. The minority argued along several lines: they feared the power of the big capitalists and that their influence would increase. They also argued that Shingū had plenty of customers, and if they lost Taiyō Gyogyō this would hardly mean any serious troubles. Finally they argued that the harbour had no space for the company. As the deal concerned the *taichigyo,* only the *gochiami* boats could have any benefit from it, and it was stressed that these boats dominated the harbour too much already. In short; the majority had taken too many liberties, and if the cooperative idea should mean anything at all, the weaker ones had to be considered. In other words; they questioned the validity of the normative rule that the co-operative should work for the benefit of all its members. The minority, which was lead by a couple of *gochiami* boat skippers and silently supported by the skippers of the smaller units, won the battle through unanimity.

Finally, the opportunities to influence the decision-making are limited by what can be called "to live up to proper behaviour". The decisions are taken in theory by a majority vote. On the other hand, unanimity is desired and valued as this "strengthens the harmony of the community". It is expected that everybody should work towards unanimous decisions, and anything else would be regarded as improper. This lays a restraint on the majority (cf. the case above) as well as the minority, as the following case will illustrate:

A11-S who is one of the active younger members, is one of the most outspoken as well. He was against the sale of sand, he was not in favour of the former chairman, in short; he is one of the most active opponents to the "establishment" represented by the committee. Then he was himself elected to represent the younger. During his time in office, the salaries of the employees were often discussed, and it was decided to increase them by ten per cent annually. He was strongly against this increase, he told me, but he felt his new position difficult. He could not oppose the others as much as he did before. He was afraid of being labelled a communist, and he was afraid of being regarded as a trouble-maker and that the others would begin to hate him. He was apparently afraid of being hated for two reasons. First, though a certain opposition is expected, too much of it is regarded as improper behaviour and feared as a threat to the group loyalty. The more responsible the position he holds, the sooner the limit of accepted opposition is reached. Secondly, as the salary implies extreme personal interests, a successful opposition could lead to enmity between him and the employees. Another member expressed the same fears. He was against the sale of sand, but he feared the poor fishermen would hate him if he voiced his opinion. He kept silent.

It is important to stress that only a few fishermen are really active at the meetings, and many follow either the opinion of their friends, or of the majority. If only a few of the influential leaders change their minds, this would therefore lead to great changes in "public opinion", as was apparent in the case involving Taiyō Gyogyō. In theory all the full members have the same opportunities to influence the decision-making, but the ability to use this greatly depends on personalities, and it is also to some extent dependent on the technology used by the members.

Despite these shortcomings, the fishermen have today obtained a certain independence in their operations. Though markets, capital and fishing rights are still limitations, they are no longer dependent on a few merchants for their survival. And they have obtained through the cooperative the possibility of allocating these limited resources among its own members. That conflicts of interest exist between the fishermen is nonetheless obvious.

CHAPTER IX

RECRUITMENT OF FISHERMEN

It should be clear from the preceding chapters that one of the important features of the fisheries in Shingū is the lack of potential recruits to the fishing vessels. The number of fishermen has been reduced from about 150 in 80-100 households to only 40 in 30 households. Why is this so? What are the limitations to growth in Shingū's fisheries? I will in this chapter make an attempt to construct a model which can depict the possibilities of growth and decline. We have to analyse the opportunities the actors have of allocating their resources, and the rewards and entailments of fishing activities compared to other activities. In other words, we cannot see the fisheries in Shingū in isolation and thus we have to relate our model also to the external world.

Economic considerations play an important role in the dispositions of the actors. However, this cannot explain why there is a higher occupational mobility among the sons of employees on the boats than among the sons of the skippers, since the net profit is divided equally. In order to understand this pattern we have to consider certain values found in Shingū, and this will be done in the latter part of this chapter.

A Model of Growth and Decline

As the fishing business stands or falls with the existence of fishing vessels, I will take the skipper with his boat and employees as the starting point. The resources that a skipper needs to exploit a maritime niche were as outlined in Chapter IV. Fishing rights which required the membership in the cooperative, knowledge and confidence in the fishing operations, capital in the form of boat and equipment, and working power of his own household as well as of his employees. These resources can be regarded as input in the fishing operations, and the catches as the output.

Before I begin the discussion of the model, it might be useful to say a few words about the access to the resources mentioned. Most of them have already been discussed at length and do not need any elaborate analysis here. The membership is as shown, either inherited by the son who succeeds as head of the household (usually the eldest son) or bought. The membership gives an opportunity to acquire fishing licences and permits the member to fish in Shingū water. Capital is made available through loans as well as through private savings. The access to labour was discussed in Chapter VI, but will be further analysed in this chapter.

The last resource to be mentioned briefly here is the resource of knowledge. Formal education is not very relevant to the situation on the boat. The exception is the education given at a senior high school in Tsuyazaki, which has fisheries as a subject on its curriculum. Only one young fisherman from Shingū has attended this school, however. Fishery colleges exist and the closest is situated in Shimonoseki. Nobody from Shingū has ever attended this college, and it is believed that they have no benefits from such schools because they are aimed at more capitalized fisheries. One of the main aims of the cooperative is to give guidance in fishing, but very little is in fact given. The agricultural cooperatives have a rather extensive educational programme, but the efforts of the fishing cooperatives seem to be more arbitrarily conducted. The staff of the cooperative attends some courses arranged by the central cooperative, and it is supposed that the given information, which mostly concerns new technologies, is spread to all the members. If anybody wants to try these out, financial support is given through the cooperative.

But why does one fisherman catch twice as much as another, both using the same technology? How does the fisherman learn the technique by which the technology is optimally exploited? This depends on personal skills which are an aggregate of knowledge mostly learnt from their fathers and through personal experience. This implies that the skipper's son becomes better prepared for the skipper status than does the son of a crew member. Exchange of information is very limited, at least if they exploit the same niches. Knowledge is closely guarded, but much can be learnt by observing the other fishermen, "spying" on each other's equipment, following after the best boat etc.

Let us assume that the skipper controls the resources mentioned, and let us follow the alternative allocations of the output from the fishing operation, with special attention to limitations and opportunities within the system (Fig. 7).

The catch can be allocated in three different ways. The fish can be consumed by the skipper's household; it can be given as part of the salary of the employees; or it can be sold at the market. It should be noted that by "consumption" I mean here the part of the catch that is taken to a sphere outside that of production, i.e. it is neither re-invested in the production nor paid as salaries. The fish consumed by the skipper's household can either be eaten by its members, or it can be given as gift. The gift can be made in an attempt to receive later favours, in which case it can be regarded as investment in influence. Frequently the fish is also given to relatives and friends on the basis of reciprocity between one article (fish) and another (rice, vegetables, cakes etc.).

The money obtained from the sale of fish can also be allocated in several ways. Usually ten per cent is spent on saving, another ten per cent on the maintenance of boat and equipment. The saving accounts make it possible to obtain credit that again can be invested in boats and equipment. Money can also be used to increase the prestige and influence of the skipper. The commission to the cooperative can, as shown in Chapter VIII, be turned into influence, and the same holds true though in a more indirect way, through religious and secular donations.[1] In this latter case the money is taken from those allocated to consumption. Furthermore, money can be invested in a wife. (In the wedding arranged in 1975 a bride price of 450,000 yen was paid, and this was regarded as normal. The bride was given a dowry that exceeded this amount).

It should be stressed that the skipper is in no way free to allocate either the catch or the money as he pleases. Normally the catch that is taken for consumption (*a* in Fig. 7), is equal to the fish given to each of the employees (*b*). If there are three persons on the boat, the fish that is not sold is usually divided by three. Furthermore, the skipper does not necessarily decide the allocation of the catch; that is done by those who sort the fish, usually members of the core and extra crews while the skipper takes care of the boat. The allocation of the money is equally restricted. The amounts the skipper can allocate to saving and maintenance

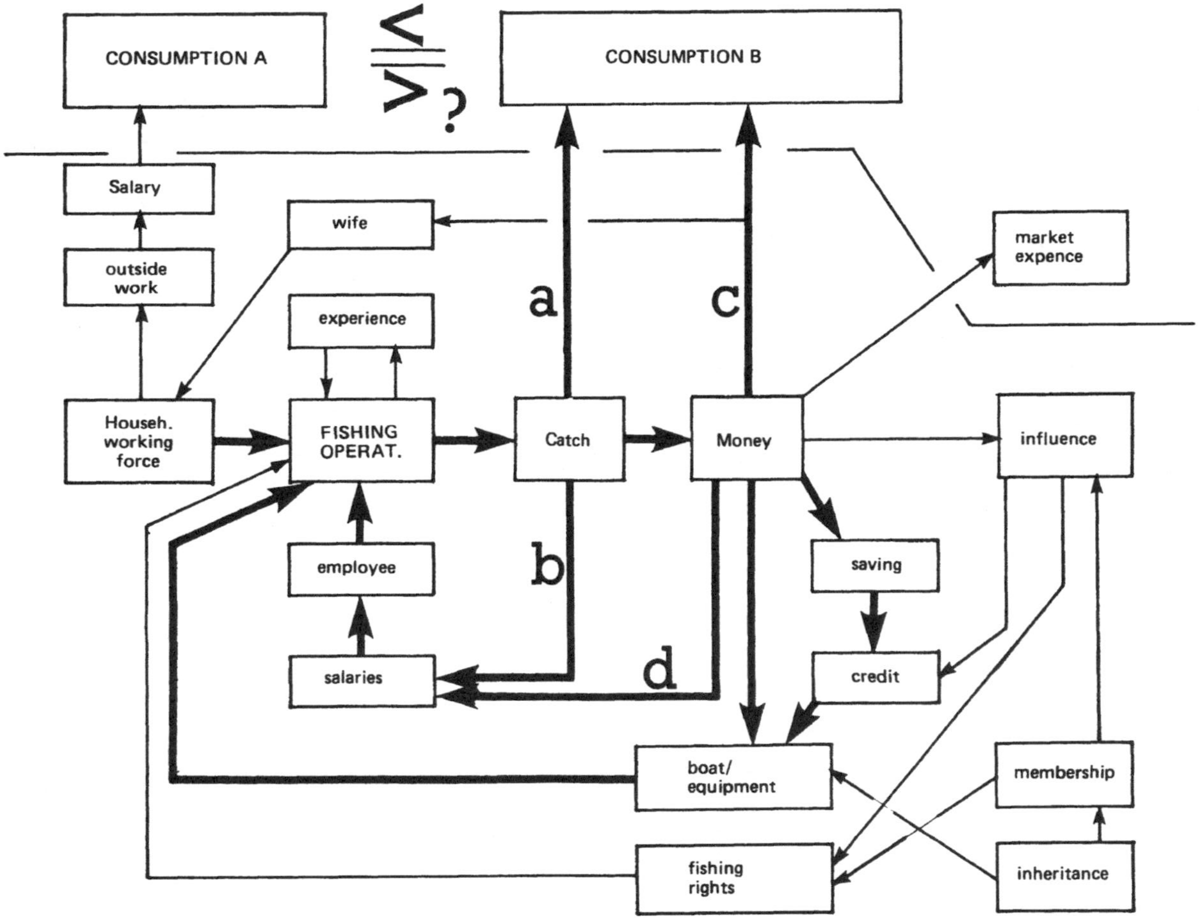

Fig. 7 The model of growth and decline

are fixed as a certain percentage of the gross sale; the same holds true for money directly invested in influence, i.e. the commission to the cooperative. Again, the money that is spent on the salary of each employee (*d*) shall be equal to that spent on consumption plus wife (*c*). In other words; the salary of each employee is exactly equal to the income of the skipper, which in turn should indicate that the standard of living of the two statuses are equal (disregarding supplementary incomes from other members of their households).

It should be clear from Fig. 7 as well as from what has been said so far, that there are several potential spirals of growth, and it is necessary to investigate these spirals in order to find the limitations of each.

1. Fishing operation → experience → fishing operation

Knowledge (including experience) does, as shown in Chapter VI, directly influence the catches, but the limit is reached when the skipper can make no further use of additional knowledge in the operation, or there is hardly more to learn. He has to utilize other technologies or exploit other niches to go further. The new knowledge required is not obtained through this spiral, but has to be obtained from outside, for instance through the cooperative.

2. Fishing operation → catch → money → salary → employees → fishing operation

The limitations are embedded in the catch which is only partially proportional to the number of employees. During the *gochiami* fisheries very little (if anything at all) is gained by employing a fourth person, and the increased catch can by no means pay this person. Though the technology can be utilized with only two persons on board, a third is appreciated, not so much because of the increased catch, but because the burden becomes less demanding. On the other hand, a reduced crew could within limits be compensated for by increased mechanization, as was the case when the engine was used to pull the *gochiami* nets.

3. Fishing operation → catch → money → saving → credit → boat/net → fishing operation

The correlation between investment in equipment on the one side and the catches on the other is not strong. A big boat of four tons does not catch significantly more than a small boat of two tons. In fact, if the boat gets too big, it might have negative implications on the catches. One skipper had to abandon *wakame* cultivation when he got a new boat which was so high that the work of cutting the seaweed became too heavy. If investments in bigger and more expensive boats should make sense, other niches have to be exploited and other technologies used. If the money on the other hand is invested in new, but not necessarily bigger, boats we get a higher turnover of equipment which again can lead to better catches. New nets decrease the necessity for dead days in order to mend the old ones, the new boat and engine save days for repair etc. This is the course taken during the last few years when most of the *gochiami* boats have been renewed. This does not imply a growth, however, but only an accelerated replacement of means of production.[2]

4. Fishing operation → catch → money → influence → credit → boat/net → fishing operation

Again we meet the same limitations as under point 3. The correlation between "money" and "influence" is moreover dependent not only on the money spent, but also on the manipulating skills of the actors. The possibilities of converting wealth into influence and then back to credit might have been better in the past when the *yakuin*, which guaranteed loans, were elected on the basis of their success in fishing. Success as a fisherman is not the important asset any more, and the *yakuin* are elected according to age and friendship. The same limitation is at work in the next potential spiral:

5. Fishing operation → catch → money → influence → fishing rights → fishing operation

The main limitation is the limited number of fishing licences. The cooperative on which the fisherman might have an influence only allocates the licences which are given by the authorities. None have two licences in the same fishery, and it is difficult to obtain a new kind of licence. Again, the possibilities for growth were present in the past when the leaders of the village could influence the village and district headmen directly.

6. Fishing operation → catch → money → wife → household labour → fishing operation

One is not allowed to have more than one wife in Japan, but concubines may be kept in secret. However, a concubine kept in this way cannot provide any labour to the fishing vessel. Moreover, there is no positive correlation between the bride price and the labour the wife provides in the fisheries. Rather the contrary is the case; the higher the bride price, the higher the social standing of the bride, and the less likely it is that she is prepared for such work. By paying a high bride price the bride-groom's family can nonetheless make a show of affluence and increase its social prestige.

In conclusion it may be stated that there are serious limitations in all the potential spirals, and growth is possible only if allocations are made simultaneously within several spirals. Investment in a bigger boat might have several implications. First, more employees may be needed though mechanization can substitute labour to a certain extent. Secondly, big boats can only be justified if new niches can be exploited and other technologies used. And then we move from limitations within the system to external restrictions.

Such growth was by no means impossible in the past. The change from simple gill nets and hook-and-line to the *gochiami* technology might be viewed as an example when the spiral of growth has been at work. The successful fisherman using simple technology managed, by allocating resources both for saving and maintenance, to change to the more profitable *gochiami*. The switch was made easier because the old boat could be used with a few modifications, and could through a limited mechanization operate with the same, or in some cases even less, personnel. The *taichigyo* fishery was easily adopted because the same technology was used though a new technique had to be learnt and capital might be needed as the catches might be temporarily smaller until new insight was obtained. The success had been tremendous, and the niche has been threatened, and quotas were imposed as a new restriction. Entrepreneurial fishermen like Hangorō and Sankichi managed to become owners of net-groups; both were in important positions of influence which at that time gave rise to cumulative growth.

Further growth today can be thought to follow two lines. The first is the one outlined above. New technologies might be introduced making it possible to exploit new niches, but this requires that some variables be kept more or less constant, particularly the access to labour (that means the new technology must not need more man-power), and licences should be equally accessible as they are today. It is doubtful if these requirements can be met unless the fishermen start breeding fish and sea-weed on a large scale, but the natural environments put a serious restraint on such efforts in Shingū. The second means of growth is to expand beyond the limits of Shingū, i.e. to go off-shore. This alternative meets again a set of restrictions. The waters around Japan are heavily exploited and regulated. A new technology of which the fishermen have no knowledge, must be learnt. The trials made outside Iki and Tsushima in the past did not encourage anybody to follow suit. Off-shore fishing will also mean a new way of life where the fishermen would be far away from home for several days, which might mean the exclusion of the women. The cooperative is furthermore not interested in such a development. If they fish in distant waters, they will also land the fish there, and this will deprive the cooperative of its own earnings.

A few fishermen have managed to grow and are today, if not rich, at least not poorer than the average Japanese. Many more have on the other hand declined, and the model can also depict this trend. If we at present assume that the economic considerations are the crucial factor in the viability of the crew, a decline can start with a poor catch. When the catches are poor, less money is allocated to saving and maintenance, and the allocations for these purposes will at the final stage be curtailed as the money is needed for the skipper's household and to keep the employees. This will in turn lead to old and rotten nets which are easily destroyed. Perhaps days are lost to repair them, and the boat which has an old engine needs repairs too. There is also less money to divide among the crew, so the wages are low. Even when some ties of loyalty exist, these will not continue for long, and the skipper might lose his best men who can get work either on a better boat or outside the fishing business. Soon he is alone on the boat, and he can do nothing but continue with old and simple equipment. Most of the boats, except the *gochiami* boats, are in this situation today. The skipper's son does not feel encouraged to take over the old and rotten equipment which is worth next to nothing, and he can get better work outside. Only two sons have taken over their fathers' boats during recent years. One has given up and joined a *gochiami* boat, the second has invested in a new engine and some equipment with money he earnt before he became a fisherman.[3] A third tried to establish himself on an old boat, but gave up and again joined his brother on the *gochiami* boats.

I have so far discussed only the possibilities of growth and decline from the skipper's point of view. Are there then no growth possibilities for an employee; how can he become a skipper? Nobody has to my knowledge managed to rise to the skipper status during the last fifty years or so. The exceptions are the cases of fission between brothers (Chapter VI) where one of the brothers has been provided with the old boat.

I suppose that the employee's failure to become a skipper lies in the economical factors, though his opportunities to acquire the necessary skills are also less than the son of a skipper's. The standard of living might be the same for the employee as for his skipper, but the latter has some undeniable advantages, as he can allocate parts of the catch to re-investment. In other words; the employee's work on the skipper's boat contributes to the latter's funds. Herein lies the most serious

Ill.38 One of the new boats in Shingū. The good catches during the last years have resulted in a high turnover of boats, and most of the *gochiami* boats have been built during the last few years.

economic barrier for an employee to acquire his own boat. The boats are expensive and last only about ten years. Traditionally the fishermen were poor, and credit seems to have always been needed to finance a boat. This credit was previously obtained through the credit associations and relatives, and it might have been advantageous to have a large group of kin which made boats of large effective *dōzoku* more viable than the others. An initial fee was compulsory in the associations and this facilitated credit, and the same is the case for the present cooperative. A savings account is needed in order to get loans. The employee cannot allocate ten per cent of the catch as the skipper does, and consequently he has to reduce his own consumption if he wishes to invest parts of his salaries in savings for a boat. In periods when the income was hardly sufficient for survival, there was no hope of allocating any of this scarce money to savings accounts or other investments.

I have discussed the model in Fig. 7 as a closed system, but certainly such a system hardly exists, particularly not in a fast changing society such as the Japanese. It is therefore necessary to investigate how external changes influence the choices of fishermen, and how recruitment to the fishing vessels is influenced by these changes, which can roughly be classified as economical and ideological changes.

The Economy of the Fishing Households

New opportunities for employment have been opened to the fishermen since the Meiji Restoration. Factories and mines developed in the Kitakyūshū region at the turn of the century, and in 1919 there were 270,000 coal miners in Fukuoka Prefecture (Mathias 1976:51). The opportunities today are much more diversified, and sons of fishermen are working as far north as Tōkyō and south as Okinawa, and in almost any kind of occupation from director in the commercial giant Sumitomo to day labourers. Work outside the fishing sector is therefore a real alternative, and economic considerations might be made. It becomes relevant to compare the standard of living – as these can be measured in consumption – between the average fisherman and the average worker. In other words; we have to compare "Consumption A" and "Consumption B" in Fig. 7. The first is simply the total wages of an employed person outside the fishing business, the latter is the sum of money and kind not re-invested in means of production and salaries to employees. In order to do so we need to introduce more quantitative data.

The alternative work available to a fisherman is naturally dependent on his education, age and former experience. A clear answer to the comparison can therefore not be expected, more so since the fishermen's catches vary considerably from person to person and from year to year. But we can hope for some tentative trends.

The difference between the boats are shown in Table 17, and several points will be made. First, the distinction between the *gochiami* boats (Group 1 in Table 17) and the others in terms of gross sale. Secondly, the other boats can be grouped into three; those operated by fishermen who are active and main earners of their households (Group 2), those who have retired or are about to do so (Group 3), and those who are employed on a *gochiami* boat most of the year, but who operate small boats during the winter (Group 4). Thirdly, four boats have joined Group 1 since 1963; three of them were formerly the very best catchers in Group 2, and the

Table 17 Gross Income as registered by the Cooperative
Fiscal years run from April through March)

Unit: 1000 yen

Management units		1963	1967	1971	1973	1975	1976	1977
	A1	681	1,949	–	–	–	–	–
	A2	1,308	2,651	3,882	3,908	7,911	7,052	12,715
	A3	1,127	2,748	3,892	3,644	6,054	6,434	11,295
GROUP 1	A4	1,208	2,777	3,776	2,923	5,973	6,646	11,637
Management	A5	1,014	2,185	4,134	2,325	3,153	4,008	9,809
units	A6	735	1,850	3,107	3,283	6,032	6,157	7,977
operating	A7	823	1,558	3,636	2,891	5,338	7,454	12,579
gochiami	A8	1,186	1,857	2,516	2,349	5,576	6,099	10,931
	A9	1,056	762	2,057	2,204	3,245	4,466	9,599
	A10	397	1,957	2,859	2,341	4,944	4,874	9,815
GROUP 2	A11	550	1,305	3,675	2,755	6,485	7,395	12,052
Units	A12	502	983	2,722	2,953	5,746	6,064	10,273
operating	A13	–	21	4,246	3,738	8,313	7,261	12,077
small nets, traps	B1	36	371	1,164	1,956	908	2,304	1,599
etc. Fishing is the	B2	78	115	729	874	1,073	1,652	2,452
main occupation	B3	217	125	152	142	175	325	–
of the household	B4	25	81	21	546[a]	668[a]	–	–
	B5	–	–	–	399	197	–	–
	C1	86	83	329	50	113	268	217
	C2	–	8	27	5	50	73	34
GROUP 3	C3	4	–	–	–	66	268	592
Units	C4	202	–	–	–	–	–	–
operating small	C5	90	14	–	–	–	–	–
nets, traps etc.	C6	61	–	–	–	–	–	–
Retired or	C7	24	2	–	–	–	–	–
semi-retired	C8	18	4	–	–	–	–	–
fishermen.	C9	5	–	–	–	–	–	–
	C10	7	3	–	–	–	–	–
	C11	15	–	–	–	–	–	–
	C12	5	–	–	–	–	–	–
GROUP 4	D1	303[b]	519[b]	554	279	405	364	1,102[b]
winter only	D2	121	67	21	–	56	63	30
	D3	9	2	36	59	28	30	46
GROUP 5 groups	Ichigo-ami	6	–	–	47	–	–	–
	Nigo-ami	300	223	–	637	–	17	–
	Kitagata-ami	1,007	266	–	737	–	–	–
	Trap-net gr.	–	–	16	5,819	2,068	–	–
	gochiami gr.[c]	–	125	2,225	3,303	3,227	3,817	3,077
	other groups	95	289	44	112	591	176	1,148

Notes:
a) The third son inherited the boat, but joined a *gochiami* boat in 1976.
b) In the period 1971-1976 he worked as a core crew on boat A12.
c) During the *taichigyo* all fish caught by the *gochiami* boats are pooled and sold at Hakata market. This income is used to pay for common expenses and an excursion at the end of the season.

fourth was created by fission from his brother. These boats are examples of successful management and growth where the skippers have been able to invest in another and more profitable technology due to success in their former. Fourthly, all those who have ceased fishing in Group 3 (with one exception) have retired of old age. Fifthly, only one *gochiami* boat has stopped fishing, and this did so when the skipper died. The son who worked on the boat did not, allegedly, have the self-confidence to become the new skipper.

The average gross sale for each group of management units is given in Table 18, and I have used these to make rude estimates of the income of each group, neglecting Group 3 and Group 4.[4] These estimates are illustrated in Fig. 8 where I have included the average income for a Japanese worker during the same period.[5]

From Fig. 8 we can assume that the income of the fishermen generally has been considerably lower than for the average Japanese worker, even when we allow for a monthly income of 10,000–15,000 yen in kind which must be added to the figures given for the fisheries. Worst off are the fishermen in Group 2, who had an average monthly income of 108,342 yen in 1977. The average *gochiami* boat fared much better, and the average income of the fishermen working on these boats was not much below that of the workers' in 1977, which was an extremely good year, however.

Table 18 The Distribution of Catches between 1963 and 1977

Unit: 1000 yen

Year		Group 1	Group 2	Group 3	Group 4	Group 5
	Total	9,138	2,194	432	130	1,408
1963	No of units	9	9	10	2	
	Average	1,015	244	43	65	
	Total	20,294	3,605	31	69	912
1967	No of units	10	9	5	2	
	Average	2,029	401	6	35	
	Total	40,505	2,395	27	611	2,896
1971	No of units	12	5	1	3	
	Average	3,375	479	27	204	
	Total	35,315	3,917	55	338	10,655
1973	No of units	12	5	2	2	
	Average	2,943	783	28	169	
	Total	68,770	3,021	229	489	5,886
1975	No of units	12	5	3	3	
	Average	5,731	604	76	163	
	Total	73,910	4,281	609	457	4,010
1976	No of units	12	3	3	3	
	Average	6,159	1,427	203	152	
	Total	130,759	4,051	1,945	76	4,225
1977	No of units	12	2	4	2	
	Average	10,897	2,026	486	38	

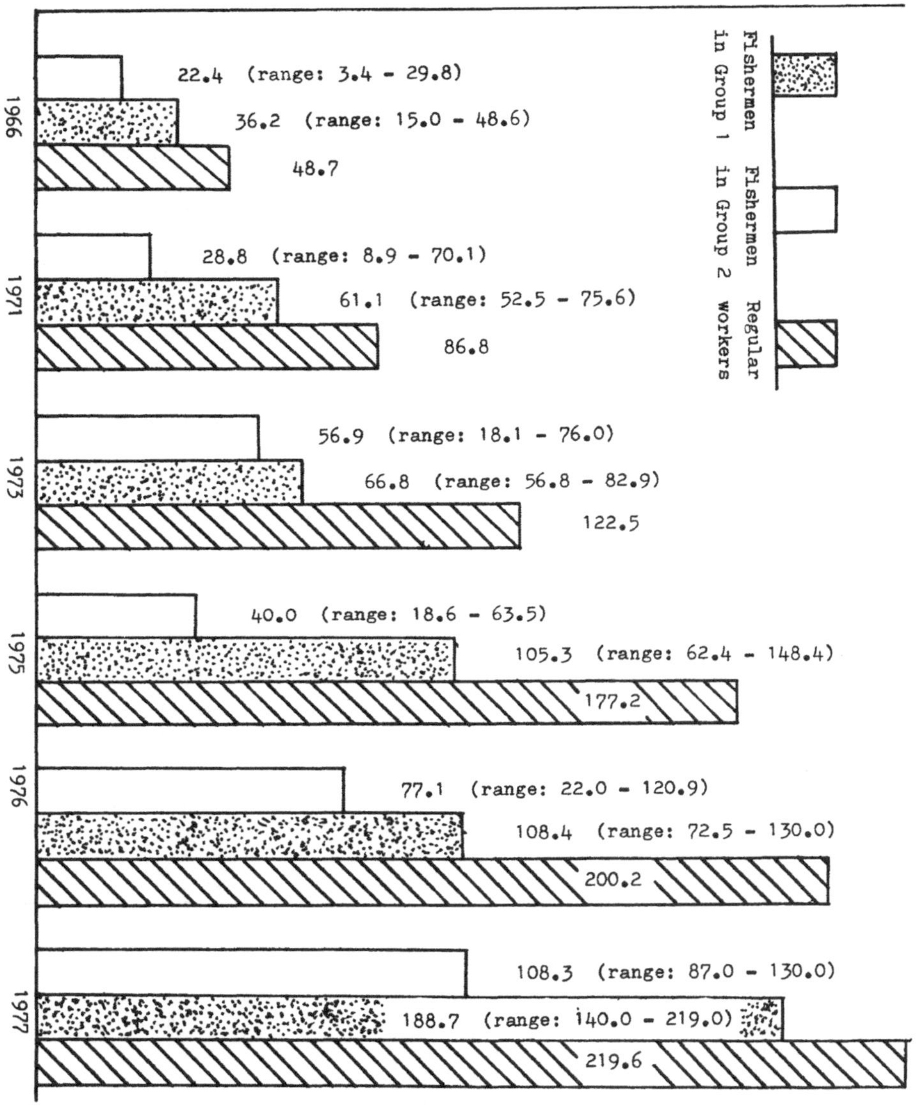

Fig.8 The average income of the fishermen in Group 1 and Group 2 compared with the average income of the regular worker (in companies with more than thirty employees).

Sources: Japan Almanac 1976:113 and *Monthly Statistics of Japan*, June 1978 (No.204): 16.

One further point should be made as we have to distinguish between the different ages of the fishermen. A fisherman has usually no education beyond the junior high school (only one has graduated from senior high school) and very little experience from other jobs. The work available to him is mostly physical work, often on a day-to-day contract, and such work could bring about 100,000 yen in monthly wages.[6] That means he will be equally well off, if not better, if he is engaged on an average *gochiami* boat. For the younger men the situation might be different. Starting with a low salary in his teens or early twenties, he will steadily climb and soon earn more than he could on the boat. For him an education and a career outside the fishing business will probably be economically rewarding.

So much about the difference in income during the recent years. But the decline in the fishing population is not of a new date. We have to compare the incomes of different groups also from the pre-war days. This is difficult because of the lack of reliable data (cf. Chapter V), but the income of the Shingū fishermen seems to have been extremely low (Table 19).

Superior skill in manipulating tax-papers cannot be the sole reason for this big difference. The natural environment of Shingū is less favourable than in the other communities. In particular Shikanoshima, Hiro and Ainoshima were blessed with good harbours attracting vessels fishing off-shore, but this does not explain fully the low figures from Shingū.

In 1935 the total value of the catches in Shingū was 30,228 yen, and there were 72 members of the cooperative. If we assume that about 100 men were active, we can estimate the average gross income to about 300 yen at the most. About thirty per cent has to be subtracted for expenses to arrive at the net income.[7] The average earnings in small factories was in 1920 about 340 yen (1.3 yen/day and 300 working days have been assumed), in 1925 it was about 400 yen (1.35 yen/day) and

Table 19 Average Income of the Fishermen in some Communities

	1917[a]		1922		1937	
	income in yen	no of fishermen	income in yen	no of fishermen	income in yen	no of fishermen
Shingū	247	81	186	150[b]	295	72
Nata	354	198	440	193	417	118
Shikanoshima	615	130	1,025	130	794	190
Ainoshima	463	108	917	106	449	195
Hakozaki	333	120	910	145	1,242	165
Hiro	446	56	892	56	–	–
Fukuma	377	53	–	–	1,452	24
Tsuyazaki	414	145	–	–	–	–
Japan (coastal)	–	–	245[c]	–	148[c]	–

Sources: Chikuhō suisan kumiai 1917, Kasuya-gun 1924, Suisan Tsūshinsha 1939, Schepers 1935

Notes: a) approximate figures only
b) probable total number of fishermen as against number of cooperative members.
c) the year 1921
d) the year 1931

about the same in 1937 (Taira 1970:34). The fisherman in Shingū clearly belonged to a low-income group.

Income and the Household Development Cycle

The household exists as shown in Chapter IV ideally of a stem family, and the changing mean household size (MHS) during the last hundred years reflects only changes in the fertility of the population, not changes in the household structure. The MHS is nonetheless essential to consider in relation to the size of the crews.

Today only four of the twelve *gochiami* boats are manned by one household (*A6, A8, A10* and *A3* which is manned by two households during the *taichigyo* season). Four boats are manned by two households (*A2, A4, A11* and *A13*) and four by three households (*A5, A7, A9* and *A12*). Thus only a minority of the households have sufficient manpower to operate their boats independently. The same seems to have been the case in 1930.

There were forty tax-payers, i.e. persons owning boats in 1930 (cf. Chapter V), and among their crew-members forty-five were also members of the skippers' households. On the majority of the boats the skippers were accompanied by less than two household members. Most of the technologies in use required on the other hand a crew of three or four (Tables 9 and 11, Chapter IV). Consequently, many of the skippers had to hire fishermen. There were registered twenty-two fishermen working as employed on the boats, and a few more were probably hired as extra crew.

If we compare the average size of the crews in 1930 and 1978, ignoring the few extra crew members, we find that the average crew has decreased from 2.68 to 2.35. During the same period the MHS declined from more than six to 4.2, and on this background it may seem surprising that the size of the crews has remained fairly stable. The skippers have managed this by bringing in their wives on the boats.

The most affluent fishing households in Shingū are, not surprisingly, those who do not need to hire additional crew members and thus do not need to divide the profit. Few households, however, manage to man their boats in this way continuously. Sooner or later they will experience a shortage in working hands due to their stages in the household developmental cycles.

The long compulsory education (nine years) coupled with the late marriage (see below) prohibits a household from contributing three generations to the household labour force at present, and we can limit our scope to the more narrow span of one or two generations.

Let us assume the father is still fishing when the children graduate from school. The daughters will usually take employment till they are married off. During these years they will contribute to the household economy, but their salaries will also be spent on personal consumption and preparations for their marriage. The sons can either take employment in a firm or become fishermen. If the father is fishing alone on a small boat, the income from this fishery is very limited indeed. This kind of technology cannot easily be adapted to two men with profit, and there is hardly money to invest in other technologies. In other words, there is no room for the sons on the boat, and they have to take employment on a *gochiami* boat or outside the business. Most have chosen the latter, which is clearly illustrated by the fate of

Group 3 in Table 17. Though they had worked in firms briefly, two have returned to fish on *gochiami* boats and then eventually have taken over their fathers' boats on their retirement or death, but one gave up after a couple of years. Those who have taken regular work outside do not feel encouraged many years later to return to their fathers' rotten boats and with much decreased income.

For the *gochiami* boats the situation is a little brighter. These boats need more man-power and several have had difficulties in getting enough crew members. Furthermore, the income is not that much below the average for Japan. And if a boat is manned by one household only, the net income of the household will be well above the average for the country. A son has taken over as skipper on six of the thirteen boats that have been engaged in *gochiami* fishing since this kind of technology was first introduced. One has succeeded in establishing himself by help from his parental house, and only one (*A1*) has not succeeded in its succession. Thus the present developmental cycle of the households can be regarded as an incentive for sons to join the *gochiami* boats, but makes it impossible to join boats not operating such nets.

If we follow the *gochiami* boat now operated by father and son, the day comes when the father has to retire, and this happens as mentioned usually before the grandson is mature. A critical period starts when the household can only provide one male worker for the boat. If he has to rely on employees, he has to divide the income, and the income can be small. The children are too small to work, and he has to rely on his wife's labour, either on the boat (the new trend described in Chapter VI) or in other occupations. If the children are still minors, this can be very difficult to accomplish, and in order to provide the family with its daily needs it can be necessary to restrict the money spent on replacement. The downward trend can thus have been started. Again, when the children begin to work, the economic problems disappear, but it might be too late to save the operation of the boat.

I have assumed that sons existed, but with the low number of children (Table 22) many couples have no sons. Previously sons were adopted, but this is increasingly difficult for two reasons: few families have sons to spare, and the young men do not want to be adopted as there are many other opportunities for house and employment which did not exist before. Though men have been adopted into some households since 1950, all of them with one exception have failed to become fishermen. The reason might be that upon adoption they were already firmly established in other occupations, or for economic and other reasons (which I will return to shortly) they did not want to be fishermen. Households with no children or only daughters can therefore no longer expect to continue in the fishing business.

We have thus seen that the economic viability of the fishing household is closely related to its developmental cycle. When several members are working on the same *gochiami* boat, the household can be rather well off, and the same holds true if the household can allocate labour in other occupations. This connection is clearly visible in Shingū today. Among the very best houses in the community are those built by fishing households who at present have several adult members each contributing to its economy. Among the poorest are those who have only one worker. In such a case social aid might be extended, or the wife is given some relief work by the township.

Ill.39 A fisherman's house. The good catches have also given some of the fishermen the opportunity to build beautiful new houses.

The Attitude towards the Fishing Occupation

We have seen that the income of the fisherman can under favourable circumstances be comparable with that of the worker, but economic considerations alone cannot explain the pattern of recruitment. Why does the son of an employee fail to become a fisherman? Another factor that greatly influences the decisions in the model is the actor's concept of value, which is a product of traditional attitudes within the fishing community and attitudes from the external world.

The fishing population in Japan has often been regarded as being very lowly people, and it never had any high status. The scanty literature in English on Japanese fishermen strengthens this impression. Norbeck in his detailed ethnography of Takashima stresses that these people turned to fishing only recently, and they did not want to be associated with the stigmatized fishing community in the neighbourhood (1954:112-114). Brameld discusses the stigmatized fishermen in Ebibara and compares them later with the *burakumin* outcaste (1968). And in *Japan's Invisible Race* references are made to the *ebune* fishing outcaste formerly living on their fishing vessels (Norbeck 1967:184-186).

It is therefore important to stress here that the fishing population in Shingū is in no way segregated or treated as a stigmatized group. The fishermen have, at least since the Meiji Restoration, participated in political affairs, and have been elected to all offices inside Shingū-ku as well as to the township. The fishermen have furthermore married outsiders for generations. Though tendencies towards village endogamy existed in the past (as it did in most Japanese rural communities), marriages between farmers and fishermen have not been rare. This is very different from the situation in Ebibara: "Prejudice against fishermen by other groups in the area has been diminishing; indeed farm girls sometimes move to Ebibara as wives of fishermen, a phenomenon described as 'unthinkable' hardly a generation ago" (Brameld 1968:59). Though prejudice against the fishing population as a group hardly exists in Shingū, such attitudes towards fishing as an occupation are present. But this is today an attitude towards an achieved status, not towards an ascribed one as is the case with the *burakumin, ebune* and other stigmatized groups.

This point needs some elaboration. The fishermen have a self-image (which is partly shared by the others) that they are dirty, hard drinkers, uncultivated, aggressive and vulgar; behaviour which is counter to the idealized Japanese behaviour of being clean, polite and well-educated in the arts of tea ceremony, flower arrangements and the like. In short, the occupation of the fisherman is not "nice looking", and they "smell". As new work opportunities are opened, many seek upward mobility and become members of the new middle class, a goal most young Japanese are striving towards. To quote Vogel (1963:9):

> The salary man's career is not a rapid and glorious rise to such great heights that it appears beyond reach, but a secure path to moderate success . . . For the vast majority of Japanese the life of the salary man seems to represent as high a standard as they can reasonably hope for.[8]

The path to the status of salary man goes through education, and most of the fishermen value education highly. The parents are eager to push their children through the senior high school at least.[9] By doing so they are well aware that the children might choose other occupations than fishing. Several of the fishermen even encourage their children not to fish, though the attitude on this question is rather divided (Table 20).

Table 20 What is your opinion about your sons' occupations?

I want them to become fishermen	2 cases
I do not want them to become fishermen	4 cases
I do not mind/they can do as they please	15 cases
NA/do not have children	4 cases

If we compare the answers to this question with the attitudes of the fathers of the present fishermen (Table 21), we might be able to say something about the change of attitude. It can seem that the attitude has changed radically in a negative direction, and this is not surprising if we consider the many new opportunities opened since the war. However, the negative attitude might have been stronger than Table 21 suggests because the two questions above were asked to present fishing households. If children were influenced by their fathers' opinions, they might have sought other occupations.[10]

Table 21 What was your father's opinion about your work?

He wanted me to become a fisherman	11 cases
He did not want me to become a fisherman	2 cases
He did not say/he let me do as I pleased	10 cases
NA	2 cases

The negative attitude cannot be directly correlated with the economic success of the fishermen. Thus the very best skipper during the last years encouraged his only son to continue at an economic university in Kitakyūshū City. He knew very well that this would not benefit his own success as a skipper, and from 1976 he had to rely more and more on his wife as extra crew member. He converted success in fishing into education for his son, and thus upward social mobility.

All the present active fishermen are sons of skippers. This fact implies that some households have declined from having a skipper to one having only a core crew member with no share in the equipment. Further it implies that during the last generation no sons of employed fishermen have themselves become fishermen. Whereas the standard of living might be equal between the skipper and his crew, other values play an active role here. The fathers' attitudes towards their sons' occupations seem to be dependent on their own satisfaction in work. Those who enjoy the work as fishermen are more favourably inclined towards their children's succession than those who fish because "it can't be helped". And those who enjoy fishing do so first of all because of the independence this occupation requires; the freedom from fixed hours and bosses, the excitement felt every day, and which by many is compared to the excitement felt in gambling.[11] The skippers are in a better position to get these values satisfied, as he decides when and where to fish. The crew members do not have the independence the skippers have, and they cannot

Ill.40 A fisherman's wedding at Isozaki shrine

influence the "gamble". On the other hand, some influence in the operation can be obtained through co-ownership, and the co-owners might get a share in the desired values.

Let me again turn to Vogel for a moment:

> The young Japanese girl hopes to marry a salary man even if his salary were low because his life is steady, he has leisure time, and she can be free of the anxieties and work connected with independent business. Independent shopkeepers, craftsmen, and farmers complain that they cannot compete with salary men in attracting desirable brides (1963:9-10).

The fishermen are no exception, and the situation for the Shingū fishermen is difficult indeed. A desired match is one where the spouses have approximately the same station in life, or the brides try to marry a little above themselves. The fishermen who have a low position from the beginning, start with a handicap. Nonetheless, they had no difficulties in obtaining brides in the past. In twenty-five marriages entered into before 1960 the average marriageable age was 26.3 years for men and 23.5 for their wives. In the seven marriages entered into between 1960 and 1965 these averages had climbed to almost 28 years for both sexes, and for the seven marriages arranged after 1965 the ages jumped to 33 years 11 months for the men and 33 years 1 month for their wives. The average age of the three unmarried fishermen was moreover 28 years in June 1978. These delayed marriages have serious consequences for the fertility rate of the fishing population (Table 22) and the developmental cycle of the household which again influences its viability.

The reason for this development is partly as suggested by Vogel. Whether the salary man has more leisure time than the fisherman is, however, doubtful, but the life offered to a wife of a salary man is secure. The same security is shared by most men employed in companies. They are employed for life (i.e. till they reach 55-60

Table 22 Marriages of fishermen since 1960

Year of marriage	Husband's age at marriage		Wife's age at marriage		Children born to the couple in (year)
	years	months	years	months	
1960	26	4	29	0	1960 & 1962
1960	29	6	24	9	1961 & 1965
1960	27	2	31	4	1971
1960	28	9	27	4	1963 & 1964
1960	25	9	26	9	1961 & 1963
1961	26	1	27	9	1962 & 1964
1961	30	10	29	7	1963 & 1964
1966	34	9	33	4	none
1968	29	0	25	4	1969 & 1973
1969	34	0	36	1	none
1974	38	2	34	1	none
1975	32	6	34	7	1976
1977	30	8	30	10	none
1978	38	5	37	7	none

years old), and on retirement they receive a lump sum. The salary is climbing steadily, and if the company is of a certain size many allowances are given: free medical care for the family, free holidays, health insurance, kindergarten and so forth. It furthermore carries prestige to be employed in a big and well-known firm. To be married to a fisherman does not carry any prestige, nor does it bring social security. The work is dangerous and demanding. It requires good health, illness is disastrous, and the income is unstable which makes housekeeping difficult.

The age at marriage has risen, and the fishermen also have to take their wives from outside the community. The latter tendency is found throughout Japan, but it is more marked among the fishermen than among the others in Shingū. The fishermen, due to their increased isolation (Chapter IV), have been more dependent on arranged marriage than previously. There are few opportunities to make friends with the opposite sex. In arranged marriages (*miai kekkon*) the parties are more conscious about the social standing of each other, a fact that works in the disfavour of the fishermen.

The fishermen today have to take the "left-overs". This does not mean that the brides are inferior in any way, but for different reasons they have failed to be married off at an early age. When they get older, their "value" as brides is diminishing, and they become accessible to the fishermen.[12] It is in this connection significant that of the fourteen couples married since 1960, the wives are the older part in seven.

Ill.41 Aged twenty-six, he is the youngest fisherman in Shingu.

Ill.42 A young fisherwoman. Brought up in Fukuoka City, she started to fish when she married a fisherman in 1977.

One further point should be made; it has become more difficult for succeeding sons to get married. Previously, social security was best found in the main household (*honke*). The branch households were usually poorer, but with the industrialization all this changed. The younger sons went to the cities to become workers, and many came to value this new independence. The young wife's position in such a household became easier. She was free from the influence of the mother-in-law, a relationship that has been analysed thoroughly by Vogel (1963), Beardsley et.al. (1969) and Masuda (1975) among others. This attitude from the young women might eventually influence the whole household structure, and several people in Shingū favour now a fission of the extended households. Such fission did, however, also occur in the past (cf. the *inkyo* institution, Chapter IV).

To become a fisherman usually implies a lower income than what he could have received in other occupations, a low social standing and insecurity. Some skippers might be prepared to accept these losses as they feel these are compensated for through his independence and excitement. The employees, on the other hand, cannot be compensated for losses in this way, and their choice of occupation will more directly be made on economic and social considerations. He also feels freer to do so, as he does not inherit any means of production as the son of a skipper does. The sentiments of the traditional occupation of the household, though not strong among the fishermen in Shingū, are more a burden on the skipper's son than that of the employee's. It can therefore be assumed that in the future all the fishermen, regardless of their status as skipper, core or extra crew member, have to be recruited from households of skippers, and perhaps co-owners.

Summary

I have in this chapter tried to focus on the problem of recruitment to the fisheries in Shingū. I took the departure from the economic point of view, and showed how the smallest management units could not provide the necessities of a household. The *gochiami* boats are more in harmony with these needs. They can provide an adequate income, in some cases above the average of the worker's household, and this technology allows the skipper's son to participate in the fisheries. Nonetheless, the social security is low, and a critical phase is reached when the father retires before his grandson can take his place. Furthermore, the status of the fishermen is low and gives no prestige, and they have difficulties therefore in getting married. This causes a low fertility rate which implies that the critical phase, mentioned above, is prolonged.

Some fishermen are willing to pay these costs. The rewards they receive are independence and excitement, values that employees do not get. The recruitment among the sons of the skippers and the employees therefore shows different patterns which will probably continue. The employees will be less attracted, and where the boat is owned by several households the incomes tend to be small, so small that additional incomes are required from other household members. Only where the boat is owned and operated by one (in rare cases two) household members does the income from fishing meet the needs of the household. During the critical phase of the developmental cycle the wife has to be prepared to join the boat, a possibility four wives exploited permanently during my stay in 1975/76.

A fifth started as a fully-fledged fisherman in 1977, a sixth in 1978, and two have participated as extra crew for several years. Skippers who do not manage to secure the labour of their sons and temporarily of their wives will, in the future, probably end their careers either as employed, or more likely, on a small management unit operated by them alone, i.e. they will join the boats of Group 2 and eventually Group 3 in Table 17.

CHAPTER X

CONCLUSION

Bailey (1969:75) has once stated that:

> We have already a variety of synonyms for strong and weak leaders. The strong leader commands: the weak leader asks for consent. The strong man has men at his disposal like instruments: the weak leader has allies. The former's political credit is high: the latter's low. The strong man has ready access to political resources: the weak leader does not. Notice that questions about a leader's control over his team are questions about the relative size of his political resources as measured against the political resources independently controlled by his supporters.

The theme of this book has been to analyse the resources leaders control and their possibilities to use these resources in action. I have in particular discussed the control of the following resources: capital, labour, fishing rights and market.

The control of these resources has undergone important changes since the Meiji Restoration. During the Tokugawa period, and even in the Meiji era, the main merchants had a profound influence on the lives of the fishermen. They could, through their influence on the village administration, allocate the fishing rights, and they could to some extent give permission to establish new net-groups. In this way they extended control of the aquatic niches. The merchants further facilitated the capital for the big nets which were required in order to exploit some of the niches such as the sardine. Because this species was also used to pay the village taxes, it became important to get all the fishermen to participate, and both physical and social mobility were restricted. The desire of the authorities to have a ready supply of *kako* for corvee labour, further strengthened this immobility. The sardine, which was the most important species in Shingū, was also an important merchandise as fertilizer for the fields, and this made the product attractive to the merchants who secured their access to this product by establishing patron-client relations with the fishermen.

The fishermen had, on the other hand, some independence in some seasons. Many of them had their own boats which they operated with small nets or hook-and-lines. During the winter when no fishing was possible, many worked in the *sake* breweries, which again strengthened the patron-client relationship with the merchants.

As long as the merchants owned the important nets, and at the same time were in important positions in the village administration which gave them access to extra-village influence, they had a firm control over the fishing population. They controlled resources which their clients did not control. The nets were, however, only of minor importance to the merchants compared to trade, *sake* brewing and land-holdings, and the actual management activities of the net-groups were delegated to caretakers or minor merchants who had some shares in the nets. The operational decisions were further delegated to *sendō* which allegedly was an inherited status of fishermen. Thus there was a hierarchic structure where each level was self-recruiting. The *sendō* had a certain influence over the fishermen in their net-groups as they possessed some special resources: they held by right of descent an office as leaders for the fishing activities; they had direct access to the caretaker of the management unit and could thus influence him; they were given economic rewards which they could invest in their relations with other fishermen; and later at least, they were given some shares in the nets.

The most important change in the organization of the fishing activities in Shingū is the development of the fishing cooperative. This organization gradually took over the functions of the merchants. The possibilities to allocate the fishing rights were first given to this cooperative, later it took over the market functions for all kinds of fish, and finally it started to facilitate loans. The ties between the merchants and the fishermen have been broken, and the final blow came when the fishermen were no longer needed in the *sake* breweries. The tax system was also revised in the beginning of the Meiji era, and taxes were to be paid in money only, not partly in kind and corvée labour as before. This fact, coupled with the industrialization which started soon afterwards, increased the possibilities for both social and physical mobility, and the fishing industry in Shingū met competition from other occupations in attracting recruits. The values the fishermen were trying to maximize were shaped more along the lines of the surrounding world. As the opportunities for other work improved and the standard of living among the ordinary workers increased, this competition became harder, and the fishermen came to control their own labour resource. This had profound implications on the vertical structures on the boats and in the net-groups.

When the merchants withdrew from the nets, a few entrepreneurs among the fishermen managed to reconstruct the net-groups. These entrepreneurs, who seem to have been of the *sendō* status, became the natural leaders of these new groups, but the structure of these was nevertheless different from the old one. First, the leaders (now called *yakuin*) were not supported in the exercise of authority from external forces as before. Secondly, many of the fishermen became co-owners of the new nets. The resources the *yakuin* controlled were thus far less than those controlled by the *sendō* in the old net groups. Moreover, the net groups became shareholding enterprises and the leaders were elected regularly. When the initiative-takers retired, a further step was taken as the leadership became institutionalized on a rotational principle. The leadership was no longer an ascribed status. At the same time the leaders were stripped of all prestige and financial compensation. All shareholders had the same resources and the leader's control of the net-group became very weak.

The leadership of the net-groups cannot be discussed without considering the relative importance of the groups in the past and today. In Chapter V, I focused on the different technologies used during the last hundred years or so. Both the niches exploited and the technologies used have changed dramatically. The niches exploited today do not necessitate big net-groups as the sardine did. The groups have little importance at present, and consequently the leadership is rotated among the shareholders without any thought of their personal abilities. Where the leadership carries more importance, as in the sardine net-groups in the past and the *taichigyo* fisheries at present, personal competence is considered, and elections are performed, but re-election is prohibited to avoid leadership being monopolized.

The decline in the fishing population has probably improved the remainder's possibilities to survive as fishermen, as it reduces the danger of overfishing within

Shingū waters. On the other hand, this decline in the number of fishermen has brought some skippers into serious trouble in acquiring crew members to their boats, and the role performance on the boats has consequently changed. Other work opportunities exist, and the skippers have to compete with these. The skipper finds it more and more difficult to face the competition, and he does not control the resource of the employee's labour as he did before. The skippers behave accordingly: the size of the crews has been reduced, they have been inclined to rely more on the labour of their own households, and have thus taken aboard their wives, which was taboo only a few years ago. Moreover, they might be forced to cooperate with an agnatic kin, who often possesses the same resources as themselves. We therefore find tendencies to more egalitarian crew also on the boats today.

We have thus a situation where the leaders have lost control over many of the resources they formerly had. The merchants do not control any resources at present, and care has been taken so nobody can monopolize these resources. Furthermore, the skippers have lost the control over the labour force. Consequently, we have got leaders that have only limited control over their teams. This development has taken place in the middle of a fast capitalization of the fisheries. It can seem surprising that this has lead to egalitarization, and I would like, for the purpose of comparison, to turn the attention to two other societies; the first in Malaysia (Firth 1966) and the second in Norway (Wadel 1972). Incidentally, the fishermen analysed by Firth and Wadel used purse seines, as did the Shingū fishermen in the past.

In Malaysia the introduction of the purse seine has led to centralization of the capital. A class of absentee net-owners has developed, and the economical differences among the fishermen have increased. The fishermen had no alternative ways of acquiring the nets, and they had only limited opportunities for alternative work. In Norway the centralization of the capital within the herring fisheries has not taken place, and Wadel concludes that the fishermen have remained as the owners partly because several "privileges" are extended to the fishermen which restrict the entry of non-fishermen into the business, which facilitate cheap loans through the State Fishing Bank, and a centralized marketing system. Similar factors have restricted centralization of the capital in Shingū. But as Wadel points out, "laws might be necessary but not sufficient to maintain a certain distribution of ownership" (Wadel 1972:118). There are several other factors restricting centralization: the pattern of inheritance making accumulation of wealth difficult, the behaviour of the species caught in Shingū today making big net-groups unnecessary, the egalitarian principle found within most sectors of the community today and so on.

Two factors I want to mention specifically here. The first is the merchants' withdrawal from the fisheries. The initiative to dissolve Toi-ami and Yorosu-ami might have come from the merchants themselves. The owner of Toi-ami had serious economic difficulties, and the catches of the fishermen seem to be low compared to the fishermen on the other fishing communities in Kasuya County. The annual income derived from Yorosu-ami was only a fraction of what Yorosuya earned from *sake* brewing and land-holdings. The second factor is the pattern of recruitment to the fishing industry. Unlike Malaysia, the fishermen in Shingū have many work opportunities open, and they could during the heyday of the Japanese "miracle" easily find employment. This has caused an occupational mobility among the propertyless fishermen, and today the fishermen can be regarded as a homogeneous group. The resources they control are the same, and their possibilities of controlling each other are limited indeed.

APPENDIX I

THE FISHING HOUSEHOLDS

Household head	Active Fishermen	No of fishermen	No of persons in household	No of generations in household	No of persons working outside fishing	who work outside the fishing business
A2–S	Hu, (Wi, 1S)	1	4	2	2	Wi, 2D
A2–C	Hu	1	3	2	0	
A3–S	Hu, YoBr	2	8	3	3	Mo, YoBr, YoSi
A4–S	Hu, (Wi)	1	5	3	2	Wi, 4D
A4–C	Hu, Fa (C2–S)	2	6	3	2	Mo, Wi
A5–S	Hu	1	3	2	1	Wi
A5–C47	Hu	1	6	3	1	Wi
A5–C53	Hu	1	3	2	2	Wi, 1D
A6–S	Hu, Fa, Wi	3	5	3	0	
A7–S	Hu	1	4	2	3	Wi, 2S, 1D
A7–C	Hu	1	4	3	0	
A8–S	Hu, Wi, 2S	3	4	2	1	1D
A9–S	Hu	1	6	3	2	Wi, 2S
A9–C	Hu	1	7	3	4	Wi, 2S, 4S, 5S
A9–E	Hu	1	3	2	0	
A10–S	Hu, Fa, Wi	3	6	3	1	Mo
A11–S	Hu	1	2	1	1	Wi
A11–E	(Hu), 2S, 3S, 5S	3	6	3	0	
A12–S	Hu	1	4	2	1	YoSi
A12–C62	Hu	1	6	3	2	2S, 4D
A12–C65	himself	1	2	2	1	3S
A13–S	Hu, Wi	2	4	2	0	Wi, 2D, 3D
A13–C61	Hu	1	4	2	3	
B1–S	himself	1	1	1	0	
B2–S	himself	1	3	2	2	Mo, ElSi
B3–S	Hu, (Wi)	1	3	2	1	1D
B4–S	Hu	1	2	1	1	Wi
B5–S	Hu	1	4	2	1	Wi
C1–S	Hu, (Wi)	1	5	2	3	Wi, 1S, 4D
C2–E55	Hu	1	3	2	1	Wi

Hu: Husband
Wi: Wife
Mo: Mother
Fa: Father

1S, 2S etc.: first, second son etc.
1D, 2D etc.: first, second daughter etc.
YoBr: Younger brother
YoSi: Younger sister
ElSi: Elder sister
himself: unmarried at present

Note: Persons in parenthesis work sometimes as extra crew.

APPENDIX II

MERCHANT HOUSES IN SHINGŪ-URA

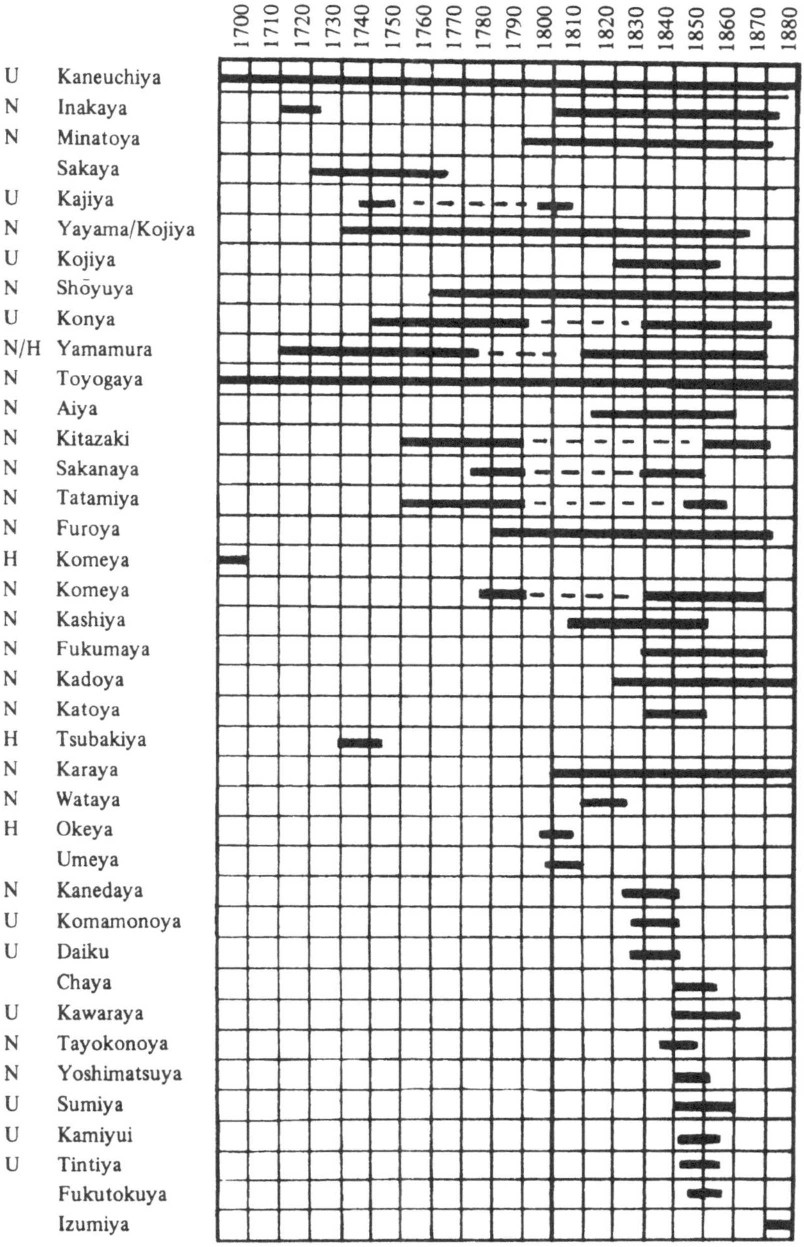

Note: H: Honmachi N: Nakamachi U: Uramachi

APPENDIX III

THE CATCHES IN SHINGŪ IN 1950 (Unit: 1000 yen)

	Jan	Febr	March	Apr	May	June	July	Aug	Sept	Oct	Nov	Dec	Total
octopus	1.1	1.5	2.0	0.3	–	–	–	–	–	–	–	–	4.9
stingray (*akaei; Dasyatis akajei*)	7.4	17.3	25.0	–	15.0	–	–	–	–	–	–	–	64.7
black porgy	1.6	–	–	1.5	16.4	–	–	–	–	–	0.2	–	19.7
sardine	9.9	–	–	–	–	–	17.1	47.0	16.0	51.1	244.6	147.1	532.8
rockfish	3.5	2.5	1.0	–	–	–	–	–	–	–	–	–	7.0
sea bream	1.9	–	–	1.6	386.4	594.1	49.7	195.1	258.8	83.5	9.3	–	1580.4
turban shell	5.2	0.7	0.5	0.1	–	–	–	–	–	–	–	–	6.5
abalone	2.0	0.4	0.2	–	–	–	–	–	–	–	–	–	2.6
squid	–	–	–	51.5	69.9	104.9	60.8	61.4	4.2	–	–	–	352.7
flathead (*kochi; Platycephalus indicus*)	–	–	–	–	17.0	–	–	–	13.7	–	–	–	30.7
flatfish	–	–	–	–	10.0	–	–	–	12.3	4.6	0.6	–	27.5
horse mackerel	–	–	–	–	–	23.0	27.0	32.5	20.4	167.3	–	–	270.2
mackerel	–	–	–	–	–	24.8	–	34.0	82.3	168.2	2.1	–	311.4
sea bass (*suzuki; Lateolabrax japonicus*)	–	–	–	0.5	30.0	–	13.1	6.0	–	–	–	–	49.6
pig fish	–	–	–	–	–	65.5	69.6	74.0	87.6	23.9	–	–	320.6
kanpachi	–	–	–	–	–	21.6	144.4	–	–	–	–	–	166.0
yellowtail	–	–	–	–	–	51.0	5.7	21.0	–	4.3	–	–	82.0
small sea bream	–	–	–	–	–	–	–	–	19.5	–	1.4	–	20.9
barracuda	–	–	–	–	–	–	–	–	15.3	6.6	5.2	–	27.1
other	4.9	4.2	15.4	41.3	55.1	38.8	37.0	65.9	45.2	16.0	0.8	–	324.6
total	37.5	26.6	44.1	96.8	599.8	923.7	424.4	536.9	575.3	525.5	264.2	147.1	4201.9

APPENDIX IV

THE RELATIVE VALUE OF THE MONTHLY CATCHES IN SHINGŪ

	1929	1940	1950	1976
January	3.56%	0.10	0.89	1.42
February	0.51	0.16	0.63	1.26
March	0.39	1.89	1.05	6.56
April	5.29	7.65	2.30	6.82
May	13.65	20.80	14.27	13.58
June	19.31	12.25	21.99	13.58
July	19.27	12.97	10.10	17.60
August	10.36	11.97	12.78	7.02
September	12.05	11.63	13.69	7.43
October	6.88	15.38	12.50	13.64
November	4.67	4.68	6.29	5.49
December	4.05	0.51	3.51	5.61
Total	99.99	99.99	100.00	100.01

The Relative Value of the Monthly Catches in Average for 1929, 1940, 1950 and 1976:

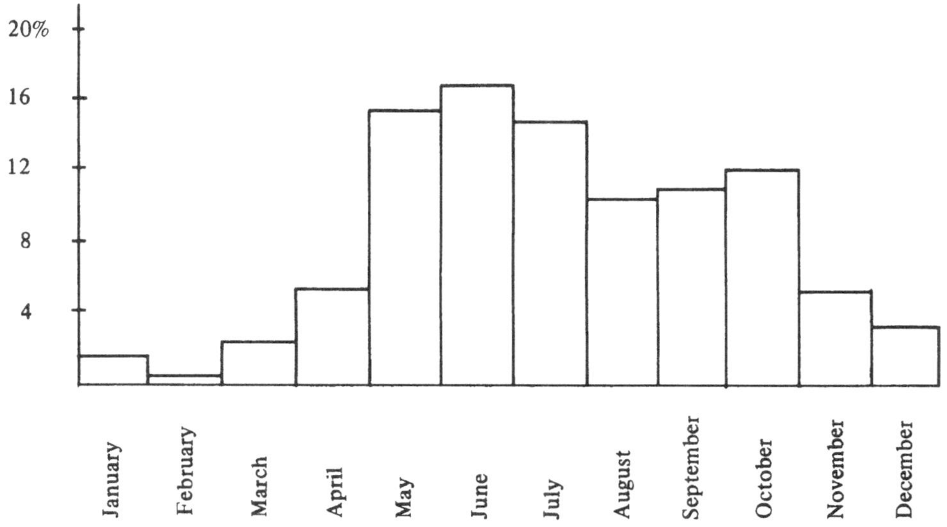

NOTES

Note to Chapter I

1. Borgstrom (1964:107) reports that the coastal fisheries only amount to 18% of the total catches though they engage 86% of the fishing enterprises. In Fukuoka Prefecture where Shingū is located, 1.2% of the boats (those over 100 tons) took in 1975 as much as 62.8% of the catches calculated in value. Calculated in tons, the percentage will be still higher (Fukuoka-ken 1975).

Notes to Chapter II

1. These formalized neighbourhood groups should not be confused with the informal group of neighbours (*tonari*) where the neighbours are those living next door. In Shingū one's neighbours are those living in the houses on either side plus the three houses across the street. Each house thus has different neighbours, and in Shingū the *tonari* is structurally of little importance though strong emotional ties may exist. When Johnson (1976:130) writes that these *tonari* "served as the first level of mutual aid" in cases of weddings and funerals, this is then not the case in Shingū (cf. Chapter IV). See also Cornell (1963).

2. Frequently the people vote not according to their political consciousness, but rather as a response to social obligations (Bernier 1975:31-32). This is particularly the case in local elections. In Shingū 57.7% answered that they voted as they did in the last election to the town council because they knew the candidate personally or were asked by a friend to vote for him; 30.9% answered that they shared the views of the candidate, his programme was good or that he had done well before. In the election to the prefecture the corresponding percentages were 20.7 and 62.5. See also Curtis (1971).

3. "Red Tide" (*akashio*) is probably caused by an unusual increase in plankton which feed on industrial wastes. The water is coloured red by the algae and it can be spread for many kilometres. The occurrence of red tide is increasing; the most serious incident so far occurred in August 1972 when "Some 14 million artificially bred young yellowtail were killed by the red tide off the four coastal prefectures of Kagawa, Tokushima, Hyogo and Okayama. An estimated 5,520,000 such fish died off Kagawa alone, causing damage totalling more than 2,800 million yen." (*Mainichi Daily News,* 11 July 1975).

4. Minamata was a small fishing community until 1908 when Nippon Chisso Corporation started operations. The city grew and prospered, but already in 1925 the fishermen received small indemnities for damage to their fishing grounds. In the beginning of the 1950s Nippon Chisso increased its production of acetaldehyde which started in 1932. Soon afterwards cats and other animals began to behave strangely. The first human attack was reported in 1953, and the disease spread rapidly. People became mentally and physically handicapped and many died. The scientists soon suspected the factory, but several reports were ignored and withheld. The fish was considered poisonous, however, and was not allowed for sale. Many fishermen lost their source of income, and this caused them to eat more of the dangerous fish. Finally the connection between the disease and mercury wastes from the factory was proved, and Nippon Chisso had to pay huge compensations. The case is still in the courts; in the meantime *Mainichi Daily News* (9 November 1975) in a small article could relate that "The total number of 'Minamata disease' victims in Kyushu increased to 760 by Thursday with the recognition of 13 new

sufferers of the disease in Kumamoto Prefecture. A total of 899 Minamata disease victims have actually been recognized – 791 in Kumamoto Prefecture and 108 in Kagoshima Prefecture – but 130 have since died." The number of victims is probably much higher, but many have not yet been recognized as such. See Smith & Smith 1975 and Gunnarsson 1971.

5. The species mentioned here are seldom used as food by the fishermen themselves, with the exception of some molluscs like octopus and squid. Most of the fishermen eat fish daily; they prefer fish to any other food, and in case of bad weather they buy fish from one of the three fish shops in Shingū. Usually they eat fish of commercially little value.

Notes to Chapter III

1. Buddhism was also brought to Japan from China and Korea, and Northern Kyūshū served as a gateway. In 805 the monk Saichō (better known by his posthumous name Dengyō Daishi) returned to Japan from China. According to an old legend he landed on the beach just east of Shingū, and he threw a stick to determine where to go. He followed the direction given by the stick and met a hunter who invited him to stay overnight. The hunter's hut was located in Kaminofu. Saichō spent a month there, and he improved the living quarters, dug a well which is still regarded as sacred, and he lit a fire that is allegedly still burning. Later upon his return to the capital, he established the Tendai sect with headquarters at Enryaku-ji on Mt. Hiei overlooking Kyōto.

The house in Kaminofu is allegedly still kept by the same family (the forty-third generation at present) and is called Sennenya (lit. the house of a thousand years). Every year on 13 April a ceremony is held at the house; on this occasion (and never otherwise) is an image of Buddha made by Saichō taken out and displayed to the public. The family feels sure the house will prosper as long as they have the well, the fire and the image.

2. Among the more important battles in the neighbourhood of Tachibana Castle is the battle of Tatara-no-hama in 1336 when Ashikaga Takauji launched a heavy attack on the supporters of the Southern Court, notably Kikuchi Taketoshi. Takauji received support from the Shōni who Kikuchi had driven from Dazaifu the same year, the Shimazu and Ōtomo.

Tatara-no-hama became again the battlefield in the fights between Ōtomo Sōrin and Mōri Motonari and his kinsman Kobayakawa Takakage, and in 1568 Tachibana Akiyoshi sided with Mōri against his overlord Sōrin. Akiyoshi was defeated, however, and he committed suicide. In 1569 Mōri and Kobayakawa managed to take Tachibana Castle, but Sōrin's forces led by Betsugu Akitsura (alias Tachibana Michiyuki) retook the castle the following year.

Finally, when Shimazu moved north in 1585 Akitsura fought against them, but was killed and his son Tachibana Muneshige continued and defeated Shimazu. After Hideyoshi's victorious campaign against Shimazu in 1587, Hideyoshi camped for some time at Hakozaki Hachiman shrine, "and his generals camped at Hakozaki, Sukiyoshi, Tatara, and other places, covering a space of about fifty square miles." (Sansom 1974b:322). Hideyoshi among other things gave orders to rebuild the war-stricken Hakata.

Another war should not pass unnoticed when dealing with the history of Shingū. In 1274 and 1281 the Mongols sent by Kublai Khan made two attempts to conquer Japan. Both times they landed in the Hakata Bay just west of Shingū, but Ainoshima was also a battlefield, a fact the inhabitants by no means try to hide.

3. Hideyoshi gave order to a land survey where the productivity of the land in terms of rice was determined. This productivity (*kokudaka*) served as the basis for

calculating the taxes of the villages and thus the revenues of the feudal lords.
4. Ainoshima has a strategic location; the Mongol invasions have already been mentioned. During the Tokugawa period several Korean missions came to Japan, and many landed on the island (in 1624, 1636, 1643, 1655, 1682, 1711, 1719, 1748 and 1764). These missions consisted of more than 400 persons, and everybody should be accommodated on the island in addition to the hosts from Kuroda's administration. The preparations to receive these missions started two or three years beforehand. The harbour and the houses on Ainoshima needed repair, and several hundred persons were engaged in these preparations (Kitamura 1976:4). The transportation of people and equipment between Shingū and Ainoshima must have been a heavy burden on the fishermen who complained (Kaneuchi Doc.128 and 126 in op.cit.:15).

There existed also a watch post (*bansho*) on Ainoshima. The purpose of this post was to guard the coast so that no foreign vessel could illegally enter Japanese waters. In addition to this came the "daily routine" like visits of the district headman and other officials.
5. The feudal lords were from 1634 forced to spend parts of their time in the capital (the so-called *sankin kōtai* system), and they had to leave their wives as hostages in the capital when absent. The *sankin kōtai* system meant a heavy drain on the resources of the lords and as much as 80 per cent of their budgets were spent on travelling and maintenance in Edo (Henderson 1965:33). This was, as Sheldon points out (1958:18), a positive factor in the development of the commercial economy, but it increased the burden on the farmers and fishermen. See Tsukahira 1966.
6. In special cases this service could take the predominance as in a fishing village in Kagoshima Prefecture studied by Kuroda Yasuo (personal communication). There fishing was for consumption only; they earnt cash by loading and unloading cargo vessels. In other places the village could be divided into halves, one was strictly fishing, the other for port workers.
7. This is still the duty of the fishermen. All young fishermen in Shingū are members of *Kyūsaikai* which can be compared to the firemen's association (Shōbōdan) on the land (cf. Chapter IV).
8. It was customary in Shingū to pull the fishing boats up on the sandy beach during the time they did not fish. This practice continued until the boats became equipped with engines, making them too heavy. The old harbour had hardly enough space for all the boats, moreover the beach was closer to the living quarters and the equipment huts. But the words of Kaibara Ekken do obviously have some relevance (Ekkenkai 1973:439):

> On the beach of this fishing village the waves go high. Even when the sea outside is calm, the fishermen can often not go out because of the surf, and they have to wait for a day or two till the sea is calm. In this way they lose time. Again, though the weather is nice and they can fish the whole day, if there is a little wind when they return home, the surf will be too high and the boats unable to reach the beach. So they have to go to other fishing villages far away and beach their boats there. In this way, the sufferings of the fishermen were great. So due to the kindness of officers a bay was constructed under the hill to the west, stone was put upon stone and a wall was made to keep the boats inside. Then it became easy for the boats to leave and return.

The harbour and the break-water were frequently repaired, but have been kept at the same site to this day.
9. This document states that since the time of Nagamasa, i.e. since Kuroda got Chikuzen province, Shingū was supposed to provide forty-one *kako* and forty-three

boats for this service. The fishermen were moreover punished for misbehaviour in 1627 and 1658, and in 1692 the duty was to provide fifty-nine *kako* and fifty-six boats. This meant that the fishing household had to make available one *kako* and one boat on average, a fact they found unreasonable since the burden in the other fishing villages allegedly was one *kako* for each three or four households. Many households in Shingū were unable to meet this duty, and people from outside the village had to be hired. The Shingū fishermen, we are informed, were reduced to poverty, and they received economic aid (*gokyūgin*). The document is a long argument for exemption from the *kako* duty for ten years (Kaneuchi Doc. 129 in Kitamura 1976).

The effect of this application is unknown, but by 1709 the merchants were supposed to participate in the duty, which lessened the burdens of the fishermen somewhat while the merchants were given some privileges (Kitamura 1976:6).

10. This famine affected the fishermen particularly hard, as illustrated in the following figure:

The number of deaths recorded among the congregation of Sainen-ji in Shingū-ura during the years 1726-33

	Honmachi	Nakamachi	Uramachi	Total
1726-31	33	48	29	110
1732-33	125	86	38	249

The number of deaths during the six years prior to the famine indicates a fairly even distribution of the population among the three streets. The number of recorded deaths from September 1732 to September the following year show a different pattern: 122 died in Honmachi, 77 in Nakamachi and 37 in Uramachi; altogether 236 people, not including the infants, died in Shingū during that year, with more than half the deaths in Honmachi where most of the fishermen were living. The merchants were living mostly in Uramachi and the central and western parts of Nakamachi.

The famine was caused by locusts attacking the rice fields in the central parts of Japan, and it was not possible for the fishermen to buy the necessary grain due to soaring prices, partly caused by the merchants hesitating to sell in expectation of higher prices (Sheldon 1958:118). The effect of the famine became so much worse because it hit with full strength during the winter when the sea is too rough to fish.

11. Several monetary units were used during the Tokugawa period. Gold coinage was counted in *ryō* divided into four *bu*. Silver was counted in *momme* and *kan* (= 1000 *momme*), but there was no fixed exchange rate between gold and silver, and bankers *(ryōgaeya)* existed to facilitate exchange. The biggest bankers "determined every day the official exchange rate between gold and silver at Osaka, which became the basis for all business transactions and price structure for the whole nation" (Hirshmeier & Yui 1975:29). Nonetheless, after 1700 there were as a rule of thumb about sixty *momme* to a *ryō* (Wigmore 1969:113), Crawcour & Yamamura 1970). The copper coinage was counted in *mon,* and the value of this fluctuated as well. Wigmore (op.cit.:114) states that the rate of exchange for the *ryō* for a long time was 4,000 *mon*, while the exchange rate was according to Crawcour & Yamamura rather stable at 6,700 *mon* to one *ryō* between 1787 and 1854 (1970:492).

Though the consumer price index based on the prices for polished rice, soy bean paste, soy sauce, *sake* and lamp oil remained fairly stable from 1751 to 1850 (Crawcour & Yamamura op.cit.), the prices could fluctuate considerably from one year to the next. One *koku* of rice was supposed to cost about one *ryō*, but between 1695 and 1733 the price for one *koku* varied from about 35 *momme* in 1718 to about 230 *momme* in 1715 (Sansom 1946:472)! Between 1787 and 1868 the mean price was 83.405 *momme* and the standard deviation was 36.88 *momme* (Crawcour & Yamamura 1970:504n).

In Chikuzen the authorities fixed the price for dried sardine to three *bu* for one *hyō* (72 litres) in 1788 (Fukuoka-ken 1963-668).

12. The position as district headman was inheritable within the Kaneuchi family until this declined. The position as village headman in Shingū was on the other hand either rotated or elected since the headman changed frequently, at least since 1790. According to different documents, the following at least served as village headmen between 1790 and 1848: Ubei (1790), Takeuemon (1794), Jiuemon (1803), Kaneuchi Shinzaemon (1806), Takeshichi (1808), Toyogaya Kyūuemon (1812), (Konya?) Koshichi (1825). Toyogaya Kyūuemon alias Heishiro (1831), Yayama Hōsuke (1839-1843), and Kaneuchi Kichisaemon (1848).

13. Each household has to pay a monthly fee to the ward budget. Today this fee is fixed regardless of the size and economic standing of the household. Formerly, however, different categories existed. Those in the highest category had to pay more than those in the lower ones (cf. Fukutake 1967:194). Economic considerations were taken when the households were so classified.

The normal sanction for breaking the community rules was to put the household one or more categories up for a certain period of time. This sanction was probably not regarded as a severe punishment as reinforcements were necessary. In 1909 it was thus decided that any household breaking the rules would be put up three categories for three years. (Note that the household was punished, not the individual.)

Households could also be fined. In 1905 it was decided that whenever a person is sent away as a soldier, one from each household should meet at Koga railway station. "If anybody returns early, arrives too late or does not participate, he will be punished by being put one category up" (Ketsugi-roku ōaza Shingū). It should be clear that this rule was hardly in the interest of the fishermen who either got a whole day spoiled or did not return in time from the fishing grounds. Probably their children and womenfolk were sent because the rules were changed the next year. It was stressed that the participants at the station should be men over fifteen years of age as they were tired of weeping women and children. At the same time the punishment for not participating properly became fixed at 40 sen. If a woman was sent, the fine was 20 sen (op.cit.).

14. In Tosa domain thirty-six items were allowed for peddling in 1692; in 1734 twelve more were included and another five in 1746. In Okayama eleven items were allowed in 1668, seven more in 1683 and thirteen more in 1705 (Yazaki 1968:164).

15. Many people come to Shingū (and Ainoshima) during the summer to swim and fish. All the year – particularly during the week-ends – the inns are busy arranging parties both for the local people and people from outside.

16. "Shinzaemon" has frequently been used by the Kaneuchis. There were at least four persons by this name. The persons adopted this name when they ascended to the headship of the family (cf. Hirschmeier & Yui 1975:38). "Matajurō" is another such name used by this house, and it seems that while Shinzaemon was the district

headman, Matajurō was in charge of the business of Mantokuya. The exact relation between them is not known. An attempt will be made to distinguish between the different "Shinzaemon" by adding given names.

17. Much of the information about Kaneuchi in the following paragraphs I owe to Kitamura Keiko (1976 and personal communications).

18. During the Genroku era (1688-1704) there were seven cases where Mantokuya acquired land from others; in Hōei era (1704-1711) sixteen cases; in Shōtoku era (1711-1716) thirty-one cases; in Kyōho era (1716-1736) 133 cases, in Ganbun era (1736-1741) fourteen cases; in Kanpo era (1741-1744) seventeen cases; in Enkyō era (1744-1748) sixteen cases; in Kanen era (1748-1751) fifteen cases; in Hōreki era (1751-1764) twenty-one cases; in Meiwa era (1764-1772) thirteen cases; in Anei era (1772-1781) seventeen cases; in Tenmei era (1781-1789) ten cases; in Kansei era (1789-1801) fourteen cases; and three to four cases each in the eras of Kyōwa (1801-1804), Bunka (1804-1818), Bunsei (1818-1830) and Tempo (1830-1844).(Kitamura 1976:12).

19. There were in 1975 two families (unrelated) living in the house which was in very poor shape. A daughter of the last Kaneuchi has moved to Shingū to take care of the family grave, and she has built a new house in the back-yard.

20. Sainen-ji was not moved together with the village; there was no space for it in Shingū-ura. But the temple was old and there was no space (so they said) in Minato-mura for a new one. Kaneuchi therefore worked to move the temple across the river. A document from 1735 sent to the authorities argued for a relocation because the temple was eaten by ants and that the fishermen needed to go to the temple before going out to kill the fish. To kill was a sin according to Buddhism and polluting according to Shintōism.

The permission to move the temple was granted, but the construction was delayed for seven years due to bad catches probably caused partly at least by the reduced population after the famine a few years before. Kaneuchi Kichisaemon (alias Shinzaemon Kichisaburō) suggested that the fishermen should donate sardine, but the fund raising went slowly until Kichisaemon donated "most of the money".

The family's memorial tablets (*ihai*) are forever honoured by being placed next to the image of Amida Buddha inside the temple building. This is the case also in the new building completed in 1975 though the family did not contribute anything to its construction. The main hall of the temple was completed in 1747 and the bell tower was built in 1763 by money donated by Kaneuchi Dōbei.

Notes to Chapter IV

1. Fukutake (1972:43 states that in certain districts where holdings were generally small, ultimogeniture was the rule. See also Naito (1970/1) for an area where ultimogeniture was institutionalized.
2. These ages are calculated in the old Japanese manner; the child is one year old at birth, and additional years are added each New Year.
3. This difference is not universal for the fishing villages in the area, as the following table shows:

Sex ratio in the fishing villages in Kasuya County
in 1825, second month. (Male/Female)

Age	Ainoshima	Nata	Shikanoshima	Hakozaki	Hiro	Shingū
11 or older	1.08	1.35	1.13	1.11	1.01	1.25
10 or younger	0.86	[illegible]	1.11	1.20	1.13	1.36
Total	1.02	1.35	1.12	1.13	1.04	1.28
Babies	2.75	1.13	1.29	1.25	0.67	1.60
Deaths	1.00	0.78	2.00	0.50	2.00	2.00
Total population	463	978	592	426	366	666

Source: Akaishi Doc. in Komonjo o yomukai 1976:108-111.

The reason for the difference between Shingū and Nata (where the difference between the sexes is significant at the 5% level) and the other four villages, I do not know.

4. Day labour could often be found in the commercial centres, and in Yamaguchi as much as 39% were so employed (Yazaki 1968:140). A decree in Chikuzen from 1772 states that the fishermen should work as such and not take day-labour even if they could earn more (Fukuoka-ken 1963:653).
5. It would from these figures be wrong to suggest that the household pattern is changing from stem to nuclear families. The farming and fishing households are mostly old households established several generations ago, while urban households are often newly established, and have thus not yet developed the characteristics of a stem family. Johnson (1964) thus convincingly argues that the industrialization in Japan has not led to the abolishment of the stem family as the ideal household.
6. In 1975 Shingū had a budget balancing at about three million yen. The main income was the community fee or *kugin* paid monthly by each member household, and this amounted to 1.7 million yen (600 yen a month per household). Another 637,500 yen was earned as rent for the use of the beach which is community property. The fish-processing industries located on this beach pay a nominal fee. Among the expenses were construction of a playground for the children (397,500 yen), pavement of the streets (364,000 yen) and erection of fences around ponds (121,300 yen); 584,000 yen were spent on salaries for the three elected executives. and contribution to the shrine and its priest amounted to 402,000 yen.
7. These data fit perfectly with the data provided by Morioka (1970/1), but she suggests that this is a new trend since 1946 when Shintō backing of local associations became illegal (op.cit.:39). My data do not support this view. If we compare the number of donators to the most important Shintō festival in Shingū, the *Gosengū* which is arranged every eighteenth year, with the number of contributors to the improvement of the harbour in 1933, we find the following figures:

Contributors to	Honmachi	Nakamachi	Uramachi	Total
Gosengū in 1934	80	66	37	183
harbour in 1933	72	68	53	193

The numbers for Honmachi and Uramachi need some comments as the difference is too big to be explained by fluctuations in the number of households. The contributions to the harbour were made compulsory, while donations to the festival were voluntary. Nonetheless, in Honmachi some people wanted to donate to the festival as individuals, not only as a member of a household. This was particularly the case where the households included persons outside the elementary family. We can assume that the same was valid for the other streets, making the discrepancy in Uramachi even greater. The immigration into Shingū was mostly confined to Uramachi, and to a lesser degree to Nakamachi, while Honmachi has long been full. The newcomers even before the war did not regard themselves as *ujiko* and thus did not donate to *Gosengū.*

8. The reason for this taboo was said to be the deity Funadama-sama, the guardian deity of the boats. This deity is conceived as female, and the fishermen say she will be jealous if women are allowed on board.

9. There are two rocks of importance on this site. One is the vessel for the deity Ryū-ō or Ryūjin, the Dragon King of the Sea, living in his underwater palace Ryūgū. Ryūjin is often represented with a tide-ruling precious stone *(tama)* in his hands.

The other stone Hakoze is the site where the deity Susanoo-no-mikoto landed unconscious. He was, according to the legend, awakened by a cock crowing, and he went in the direction of the sound. He was found by Shinozaki who made him the *ujigami* of the Shinozaki family. The priest at Isozaki-jinja is allegedly a descendant, and during the *Gosengū* festival (see note 7) only men with the surname Shinozaki are allowed to carry the *mikoshi* (a palanquin shrine). Since a cock saved the deity, the fishermen have traditionally not eaten chicken and eggs, a taboo that is still strong on the boats. No fishermen would dare to prepare such food to be eaten on board, and in cases of long periods of poor catches, they avoid this kind of food altogether.

10. Sainen-ji belongs to the Seizan Jōdo Sect with 131,000 adherents (Agency for Cultural Affairs 1972). This sect was established by one of Hōnen's disciples, Shōkū (1177-1247).

11. Honmachi started to be represented by three, when one representative refused to retire when his term was ended. He allegedly enjoyed the work.

Notes to Chapter V

1. In Japanese the name of a net is either formed by putting *ami* as a suffix to the name of the fish caught by that net or by adding *ami* as a suffix to a term characterizing the net. Thus *iwashi-ami* means sardine net, and *jibikiami* beach net. I will insert hyphens between the name of the fish and the net as in *iwashi-jibikiami* (sardine beach net). However, English terms will be used as much as possible.

2. Whale was not infrequently caught in this way, and several living members could recall cases. Regulations were issued to handle such occurrences during the Tokugawa period. The district headman should immediately be called and let merchants buy the whale on auction. One third of this income was the villagers', the rest should be "donated" to the authorities.

3. The large beach seine *(ōjibikiami)* was invented in the early eighteenth century. Previously the beach net was cast by one boat; one end of the rope was fixed to the

shore while the boat made a half circle casting the net. The new net was on the other hand cast by two boats. The bag was first cast, then the boats separated casting the wings of the net. By this new net the fishermen became more mobile and could go out to meet the shoals. They were also considerably larger than the smaller beach seines, which nevertheless have been used till today. (Cf. Kujukuri Research Committee 1958:20-21.)

4. In 1880 Shingū had the following taxable nets; the taxes to be paid for the nets are mentioned in parenthesis. Sardine net (2 yen), sea bream net (1.50 yen), beach net (0.60 yen), and the following nets each paying the lowest tax of 0.20 yen: sand lance net, yellowtail net, *konoshiro* net, ordinary gill nets, *tatsukuri* net, squid net and dragnet (Ito 1949:3). See also Fukuoka-ken (1878) for a description of the wide variety of nets used at the time.

5. The following figures based on information found in the statistical year-books (Tōkeisho) published by Fukuoka-ken, 1902, 1905, 1910, 1925, show the decline of the sand lance fisheries in the Taishō era (1912-1926):

Catches of sand lance in Fukuoka Prefecture 1897-1925

	1897	1900	1905	1910	1921	1925
Catches in tons	526	1627	2994	1136	684	112
catches in yen	30,274	49,501	198,400	68,421	164,534	47,436
Price in yen/ton	57,55	30,42	66,27	60,23	240,55	423,54

6. The sand lance was either boiled in soy sauce and sugar *tsukudani*) or it was boiled in salty water and then dried to be used in soup. This was the work of the fishermen's housewives.

7. The catches of the sand lance in Chikuzen area of Fukuoka Prefecture did not show any dramatic decline at this time. The catches in tons were:

1955	1960	1965	1970	1975
737	2898	716	1140	1167

(Fukuoka-ken 1955:121, 1960:145, 1965:172, 1970:175, 1975:133)

The last catch of sand lance in Shingū was taken in 1975. There were two boats fishing sand lance and the total catch was 1,379 kg to a value of 55,150 yen giving a price of only 40 yen/kg; i.e. far cheaper than almost anything else. It seems that there is no profitable market for sand lance in Shingū at present.

8. The Yayama main house has contributed a series of influential fishermen from the Meiji Restoration onwards. This house was the *honke* (main household) for a small *dōzoku* that formed their own net-group (cf. Chapter VII). Yayama Gonji was allegedly the founder of the first fishing cooperative in Shingū in 1872, and he was the only fisherman among the temple representatives in 1896 (Chapter IV). Asaemon and Rinzō, his sons, were also chairmen of the cooperative as was his grandson.

The connection between this Yayama and the important merchant house with the same name in Shingū during the Tokugawa period is uncertain. The present head of this formerly fishing household claims that Gonji's father Gonuemon was village headman during the Tokugawa period and that the merchant Yayama Yauemon was the head of a branch household.

I find this latter claim hard to believe as Yauemon's household was one of the most important ones already in the eighteenth century. Yauemon's descendants claim on the other hand that the fishermen with the surname Yayama received this name as they belonged to Yauemon's net-group.

9. The catches in December used to be small, and 1929 is probably no exception. The value of the catches during the eleven first months were 29,591 yen, while another document states that the value of the total catch for the year was 30,840 yen. The difference (1,249 yen) seems a reasonable catch for December.

10. Such information exists for some of the nets in 1950.

net	used 1950	not used	total	total no. of pers.
sand lance net (*kanagi-ami*)	5	25	30	130
sea bream offshore net (*okitoriami*)	2	1	3	70
sea bream beach net (*jikogiami*)	0	5	5	120
two boats *gochiami* (*nisō gochiami*)	2	5	7	70
one boat *gochiami* (*issō gochiami*)	4	4	8	24
squid basket-net (*ika-subikiami*)	10	9	19	70
yellowtail net (*yazu-sashiami*)	4	7	11	45
mullet net (*bora-makaseami*)	0	2	2	50
sardine gill-net (*iwashi-sashiami*)	6	3	9	45
nuikiriami	2	0	2	80

11. Another net that has been readily accepted is the octopus basket-net (*tako-kago*) which was first used in the fall 1977, and by December seven boats operated with this new net.

Notes to Chapter VI

1. The Shingū fishermen sell most of their fish at the big Hakata market, and the prices are not dependent only on the catches in Shingū, but on a wide area. Most of the *kisugo* is moreover sent to markets in Ōsaka and Tōkyō, and the prices are dependent on the demand at these markets as well.

The size of the market in Hakata stabilizes the prices to some degree, and the buyers know fairly well the supplies of the day before they begin to bid. The situation is therefore not at all so fluctuating as Firth (1966) has so vividly described from Malaysia. Nonetheless, the prices can fluctuate rapidly, and the prices obtained for *mame* illustrate this: On Saturday 14 June 1975 one box (5 kgs) of *mame* fetched 1,600 yen; on Monday the price was down to 800 yen. On Monday 30 June the box fetched only 600 yen, the lowest I recorded for the year, while the following Monday the price was 1,800 yen which was the highest I recorded.

2. The relative higher marketing expenses for a low-priced fish can be illustrated by the difference between *kisugo* and *mame*. During June 1975 which was a very good month for *gochiami*, the catches for this technology were:

	Value		Number of boxes	
Species	in yen	in %	No.	in %
kisugo	3,727,600	61.0	961	32.0
mame	1,807,000	29.6	1,594	53.0
other	575,300	9.4	452	15.0
Total	6,109,900	100	3,007	100

Note: There were six kg of *kisugo* in one box, while a box of *mame* contained five kg. A box of "other" is not sold according to weight.

When a box fetched a price of 500 yen or less, the fishing cooperative encouraged the fishermen not to sell, but process the fish for home consumption. However, the fishermen are tired after a day on the sea, and sometimes boxes are sold for as little as 300 yen which gives no surplus to divide among the crewmen, though the owner anyway will be able to deduct the ten per cent for the boats and thus benefit in economic terms.

3. To understand the behaviour of the fishermen, it is necessary to describe the market situation, though it may seem tedious. The fish is sold either by quantity or quality. The fish that is sold according to quantity is scaled by the employees of the cooperative. Cheating by weight is seldom attempted since, if discovered, Shingū will lose reputation and fetch lower prices in the future. The boxes sold according to quality are not scaled, but the buyers can investigate their contents.

The *kisugo* varies greatly in size, and the price is higher the bigger the fish. Boxes of *kisugo* are therefore roughly sorted in the market, and fish of size A is sold in bulk as is the fish of size B etc. It is usually the smaller fish in the box that decides the classification of the whole box. It becomes therefore important for the fishermen to sort strategically. Some try to place the bigger fish on the top of the box, but the boxes are shallow. A big catch, on the other hand, makes a finer gradation possible, and a big catch tends to imply a better price for the fish. Thus, one of the better boats in June 1975 received 4,450 yen in average for the boxes of *kisugo* while the poorest boat received only 2,377 yen on average.

If the big *kisugo* did not fill one box of six kg, two possibilities were open; either to fill it with smaller fish and thus reduce the price of the big ones, or to sell it as it was according to quality, but then pay relatively more for ice, box, storage and transport to the market.

By 1978 the contents of the boxes were reduced from five kg for *mame* and six kg for *kisugo* to four kg, making a finer grading possible. The fishermen have the impression that they are now better paid for the fish.

4. The difference between "core" and "extra" crew resembles what Bailey calls "core" and "following" in the political sphere (1969:44-49).

5. In the following discussion each person will be distinguished by a letter referring to his boat, followed by a letter indicating his position on that boat. *A3-S* will then refer to the skipper on boat *A3* (Fig. 1). Where there are more than one person with the same status on the boat, his age will be added, thus *A8-C23* means the member of *A3*'s core crew whose age is 23.

6. The same is true when father and son fish together; only one of them (usually the father) has the right to participate and vote at the meetings in the cooperative. This causes some conflicts that will be discussed in Chapter VIII.

7. This is in contrast to the practice in the communal affairs where the wives may represent the household heads as well as the *kumichō* at meetings (Chapter IV).

Notes to Chapter VII

1. One of the crew members complained because the expenses for the boat were calculated twice; first as 0.5 share and then by ten per cent of the gross income. In the example above the boat-owner received 116,200 + 127,555 = 243,755 yen for his boat's participation.

2. These numbers need some comments. They include only boats and nets that received loans from the *tame-gin* fund, and might thus not be exhaustive. On the other hand, the number of boats might be exceptionally high as Shingu was hit by two typhoons in 1828. Consequently loans were extended to eighteen boats that year alone. On average each boat received about 300 *momme* in loans, but the nets

received much more. The six nets that received financial aid were (Akaishi Doc. in Komonjo o yomukai 1976):

1823: Kiyosuke borrowed 1 *kan* to repair *tatsukuri* net.
1824: Masahei borrowed 1 *kan* 500 *momme* to repair sardine net.
1825: Kyūsuke borrowed 1 *kan* to repair *tatsukuri* net.
1827: Yōhei borrowed 2 *kan* to repair mullet net.
1829: Kyūhachi, Busuke and Kinpei borrowed 1 *kan* 500 *momme* to repair sea bream net.
1830: Tōbei and Tachibei borrowed 2 *kan* to repair *tatsukuri* net.

3. Sardine was a very important item in the Tokugawa period, and there were wholesale dealers (*tonya*) for salted sardines in Ōsaka already during the first years of the period (Sheldon 1958:46). In 1714 commercial fertilizer was the third largest item in that city (Smith 1959:83). The same author writes that:

Commercial fertilizers – chiefly fish, oil cakes and night soil collected in towns and cities – were perhaps the most important of all innovations in raising yields. Not that these fertilizers were unknown before the Tokugawa period, but they were confined to villages located on the outskirts of towns or near the seacost where the frontiers of farming and fishing met (ibid.92).

In Shingū dried sardine (*tatsukuri*) was used in the fields in Minato-mura and Shimonofu-mura in 1790 (Isozakigū Doc.4-2 in Kitamura 1976:20).

4. Each neighbourhood group has a name derived from the name of the street plus a number. "Honmachi-5" refers then to the group no. 5 along Honmachi.

5. The decline in the fishing population was mentioned as one reason why the sardine fisheries ended. I do not believe, however, that this was the decisive reason. If the sardine fisheries were profitable, the decline might have been less; secondly, mergers of the existing net-groups could have taken place, as happened when Sankichi-ami and Kitagata-ami merged. Such mergers were not discussed.

6. During the *taichigyo* fishery conflict was avoided by an agreement that if they caught less than 500 fry per day and boat, they should switch to individual fishing. The profit of 500 fish was regarded as equivalent to the average catch for individual *kisugo* fishery. The best crews wanted nonetheless to lift this limit as they felt they were losing.

7. Both Firth (1966:162-164) and Kujukuri Research Committee (1958:110-114) regard such loans as seasonal advances, and they are given to secure crewmen for the net-groups. The borrower was not pressed to return the money as long as he remained in the group. The loans are instrumental and the lender might purposely not ask for the return because "If neither side is owing then the bond between them is comparatively fragile. But if accounts are not squared, then the relationship is maintained by the virtue of 'the shadow of indebtedness', and there will have to be further occasions of association, perhaps as occasions of further payment." (Sahlins 1974:222)

Notes to Chapter VIII

1. For more detailed accounts on how this system worked, see in particular Kujukuri Research Committee (1958) and Chang (1971).

2. The fishermen of Shingū frequently cross into the waters of the adjacent fishing communities, particularly into Fukuma's. This is to a certain extent tolerated by that fishing cooperative as the Fukuma fishermen do not exploit the same resources (*kisugo*) as the Shingū people. The trespassing should not be too frequent, however,

and several of the fishermen felt there had been too much lately. The Shingū cooperative has received complaints. Such complaints are usually followed by a gift of some bottles of *sake* and apologies. But the message is also passed on to the fishermen, who restrain themselves for some time. The sanctions are mild indeed, unless trespassing is discovered by patrol-boats operated by the maritime police. Fishing boats caught by one of these will be fined, the fishing licence endorsed, and they are forbidden to fish for some days. Too many endorsements means loss of licence. The cooperatives seldom report incidents of this kind, obviously reasoning "this time me, next time you". It is only when the balance between "this time me" and "next time you" is severely disturbed that the message and complaint are communicated to the other cooperative.

3. The cooperative allocates its territory to different purposes, and the conflicts between the fishermen do occasionally develop. When a section of the territory is allocated to the *wakame* cultivation, this implies that hardly any other technology can be used in that section between November and May. *Gochiami* is impossible, and traps for ivory shell, octopus and squid can be handled only with difficulties. A similar conflict arose when four fishermen wanted to try the large trap-nets. These trap-nets occupied large areas, and the plan could only be accepted if all the fishermen were given the opportunity to participate. Practically everybody became members of the trap-net group, and everybody could receive financial assistance and licences to cultivate *wakame*.

4. There exists an agreement between Shingū, Shikanoshima, Nata and Hiro about fishing licences in the area, and no further permission is needed to operate traditional nets like the beach seine, several kinds of gill nets, sand lance nets, etc. Newer technologies are on the other hand not covered by this agreement.

5. "The word *kō* derives from *kōgi,* meaning to expound to the Buddhist sutras, from which it gradually came to mean the group of people who gathered together to listen to such expositions" (Fukutake 1972:106). The *kō* were originally religious in character, and there is still one in Shingū, the Kannonkō (Chapter IV).

The *kō* spread, however, to the more secular world, and became important credit associations all over Japan. Some were aimed at a special purpose, as the *gyosenkō.* There were before World War II also a few general *kō,* but it was quite expensive to participate, and only the richest of the fishermen were members here. One complained that the other members had better education than the fishermen, and were therefore more clever, which implied that the fishermen were badly cheated. For more extensive discussions on the *kō* institution, see Embree (1939), Fukutake (1972) and Nakane (1967).

6. The Sakugorō-kō and Heikichi-kō got their names from the boat-builders. A few fishermen denied that Heikichi who came from Nata, organized any *kō,* but Sakugorō-kō was with its more than thirty members at any rate the biggest.

7. Relative age is extremely important in these elections. The *yakuin* are of two categories; five *riji* and two *kanji* who are supposed to be younger fishermen. The chairman is elected by the *yakuin* and is usually the eldest. In the elections in June 1978, however, the eldest of the elected *yakuin* refused to accept his election to the committee. The second eldest became the new chairman, and a fifth *riji* had to be found. It was impossible to elect one of the young industrious fishermen since he would be younger than the two *kanji* already elected. The cooperative members instead turned to one of the semi-retired fishermen who had always been little active in the fishing cooperative.

8. The manager at that time had relatives among the *yakuin,* and he also operated a small company that extracted the sand. This double-role was later criticized, and

he withdrew from the cooperative. He made allegedly a nice profit which he invested in a petrol station.

Notes to Chapter IX

1. Most of the households donate regularly to the shrine, temple, fire brigades etc. The size of the donations are made public by wooden boards. People expect the rich and influential to donate more than the average.

2. The good catches during the last years have given most of the fishermen the opportunity to replace their boats, even when these were rather new. In the following table it will be clear that the *gochiami* boats are much newer than the others. It should also be mentioned that most of the *gochiami* boats were bought new, while most of the other boats were bought second-hand.

Boat	previous boat built	previous boat tonnage	present boat built	present boat tonnage	Comments
A2	1971	3.04	1975	4.50	
A3	1970	2.40	1973	3.78	Bought in 1978
A4	1968	1.84	1976	4.12	
A5	1968	1.76	1973	3.78	Bought in 1978
A6	1961	1.80	1974	3.08	
A7	1971	3.04	1976	3.78	
A8			1973	4.12	
A9	1962	1.73	1970	1.87	Bought in 1977
A10	sank in 1974		1975	2.90	New engine in 1977
A11	1963	1.76	1975	4.52	
A12	1965	1.87	1976	4.21	
A13	1970	1.87	1976	4.57	
B1			1962	1.28	Bought in 1974
B2			1963	1.76	Bought 1974, new engine in 1976
B3			1965	1.01	
B5			1968	1.86	Bought in 1973
C1			1962	0.88	Bought in 1972
C2			1968	0.85	
D1			1963	0.99	
D3			1960	0.79	

3. Since the end of World War II it has been common for fishermen's sons to work for some years in other occupations. This has been done when their labour on the boats was not needed at certain times due to the stage of the developmental cycle of their households and the crews. Though some of the money earned in this way found its way into the household funds, much was spent on personal consumption or savings for marriage.

4. I have arrived at these estimates in the following way. I have assumed that about forty per cent of the gross sale is spent on marketing, replacement and running expenses. The net sale is then sixty per cent of the gross sale. For Group 2 I have taken this as the income of the fishermen as there is seldom more than one person on each of these boats. In order to calculate the income of the fishermen engaged

on the *gochiami* boats, I have divided the average net sale for this kind of technology by three. Then I have added the income from participation in net-groups. I have assumed that all the fishermen in Group 1 and Group 2 have participated equally and that the membership was thirty-five. I have therefore divided the net sale of the net-groups (which I have set at 70% of the gross sale) by this number.

5. This average is the wages of regular workers in companies with more than thirty employees. The difference between big and small companies is considerable. In the first half of 1975 workers in companies with more than thirty employees received a monthly salary of 147,015 yen before bonuses were added, as compared to 113,822 in smaller companies (*Japan Almanac* 1976:111). Alternative work for the fishermen will be found mostly in the small companies. On the other hand, men receive a much higher salary (141,200 yen on average in 1973) than the women (70,800 yen), and this would bring the average male income upwards as compared to the figures in Fig. 8. (*Statistical Handbook of Japan* 1975:116). The figures should therefore be taken as indicators of trends only.

6. During the winter 1975/76 the fishermen working on construction sites received about 3,800–4,000 yen/day; the regular employees about 500 yen more.

7. The capital invested in the fisheries in 1925 was, according to a document, as follows:

74 boats, 120 yen/each	=	8,880 yen
6 sea bream nets, 250 yen/net	=	1,500 "
10 flying fish nets, 150 yen/net	=	1,500 "
20 yellowtail nets, 100 yen/net	=	2,000 "
60 *konoshiro* nets, 40 yen/net	=	2,400 "
15 squid trap-nets, 50 yen/net	=	750 "
40 squid basket-nets, 30 yen/net	=	1,200 "
20 *setsuki* nets, 20 yen/net	=	400 "
20 long-lines, 20 yen/each	=	400 "
2 sardine purse seines, 3,500 yen/net	=	7,000 "
4 large beach seines, 3,000 yen/net	=	12,000 "
Total investment		38,030 yen

This list is probably not exhaustive as several nets (*hachida* nets, sand lance nets, dragnets, gillnets) and hook-and-lines are not included (cf. Table 11).

8. The Japanese call the white collar worker *sarariiman*, and Vogel uses the term "salary man" to conceptualize such a worker.

9. The question "Which education do you think is suitable for your children" was answered thus:

		University		College		Senior		Junior	
		No	%	No	%	No	%	No	%
Fishing Households	Eldest son	8	57.1	0	–	5	35.7	0	– –
	Younger son	4	44.4	0	–	4	44.4	0	–
	Daughters	2	11.8	2	11.8	12	70.6	0	–
Non-fishing households	Eldest son	120	76.9	1	0.6	29	18.6	1	0.6
	Younger son	72	78.3	3	3.3	14	15.2	1	1.1
	Daughters	23	17.2	46	34.3	61	45.5	1	0.7

10. In a survey conducted by the fishing cooperative in 1950 all the fishermen stated that they were not satisfied with their fishing, which they found very hard for two reasons particularly: the poor harbour at that time, and the strong winds. Surely, such an attitude at that time did not encourage the children to become fishermen.

11. Many of the fishermen are fond of gambling, particularly the *pachinko,* mahjong and boat race. Some fishing households have allegedly been ruined due to gambling, and these have left the business and in some cases Shingū.

12. For an excellent description of this process, see Tanizaki's novel *The Marioka Sisters.* The third daughter, Yukiko, of a declining Ōsaka merchant house has difficulties in getting married partly because the family does not manage to adjust itself to its lower status. As the daughter gets older, the ambitions are lowered until an old bachelor who has failed in his attempts to become an architect, is accepted as a suitable bridegroom. Yukiko has at that time passed thirty-three.

GLOSSARY

Note: Japanese words that are explained in the text are not included in the following list.

ami – fishing net.
amimoto – net-owner.
Bon – a festival when the ancestors are honoured.
bugyō – magistrate, a Tokugawan administrator.
bunke – branch household.
burakumin – an "outcaste".
chidai – a sea bream. *(Evynnis japonica).*
chiku – a ward or hamlet, also called *ku* or *buraku.*
dōzoku – a kinship group consisting of *honke* and *bunke* households.
ebune – a fishing "outcaste" formerly living on his boat.
Fujinbu – the women's section of the fishing cooperative.
Fujinkai – the Women's Club.
gochiami – a seine similar to the Danish seine.
goningumi – a neighbourhood group.
Gosengū – a Shintō festival arranged every eighteen years.
gyosenkō – a credit association to finance boats.
hachidaami – a square net lifted by eight boats.
Hangorō-ami – a net-group operated by Kirishima Hangorō.
hōji – memorial service for the ancestors.
honke – main household.
honkuji – the winning lot in a credit association.
hyō – 72 litres (0.4 *koku).*
Ichigō-ami – a net-group.
ie – house, household.
inkyo – retirement from the headship of a household.
kako – corvée labour for the fishermen during the Tokugawa period.
kami – the concept of supernatural power in Shintō.
kan (or *kanme*) – here a monetary unit equal to 1000 *momme.*
Kanmata-ami – a net-group.
Kannonkō – a religious association based on the bodhisattva Kannon.
kanpachi – a fish *(Seriola purpurascens).*
katakuchi-iwashi – an anchovy *(Engraulis japonica).*
kisugo – a fish *(Sillago sihama).*
Kitagata-ami – a net-group.
kobun – a client to an *oyabun.*
kōjiya – a house producing malt, yeast and paste for soup.
koku – about 180 litres. One *koku* of rice was in theory the amount of rice consumed by one person in one year.
konoshiro – a fish *(Clupanodon punctatus).*
kuchō – the headman of the ward or hamlet.
kumichō – the headman of the neighbourhood group.
kumigashira – assistant to the village headman during Tokugawa period.
Kyūsaikai – a sub-division of the fishing cooperative the purpose of which is to rescue ships in trouble. Also a branch of the Fire Brigade.

madai – a sea bream (*Chrysophrys major*).
ma-iwashi – a sardine (*Sardinops melanosticta*).
mame – a term used locally for dead sea bream fry.
Mantokuya – the leading merchant house operated by Kaneuchi.
momme – a monetary unit.
mura – a village. During the Tokugawa period a farming village.
nakagai – a broker, middleman.
Nigo-ami – a net-group.
nisō gochiami – *Gochiami* operated by two boats.
nori – a laver (*Porphyra*).
nuikiriami – a seine.
ōaza – a unit within a village or town. Usually the same area as the village during the Tokugawa period.
on – blessings.
oyabun – a patron to a *kobun*.
ōyakuin – one of the main leaders of the net-groups.
Ryūgū-matsuri – the annual festival for the fishermen.
sake – rice-wine.
Sankichi-ami – a net-group operated by Inoue Sankichi.
seinenkai – Young Men's Club.
sendō – (1) a skipper on a boat, (2) an operational leader of the net-groups.
shinrui (or *shinseki*) – cognatic and affinal relatives.
Shōbōdan – the Fire Brigade.
Shōjokai – Young Women's Club.
shugo – provincial constables during Kamakura period.
sumō – Japanese wrestling.
taichigyo – the living fry of sea bream.
tame-gin – a fund during the Tokugawa period from which the villagers could borrow money to repair houses, nets, and boats.
tatsukuri – a small *katakuchi-iwashi* dried to be used as fertilizer.
tatsukuri-ami – net to catch *tatsukuri*.
teguriami – a small drag-net.
teichiami – a large trap-net.
Toi-ami – a net-group based on Toyomura's *sake* brewery.
tonarigumi – neighbourhood group.
tonya – operator of a market, a wholesaler.
Toyogaya – one of the main merchant houses in Shingū.
ujigami – a tutelary deity.
ujiko – person under the protection of the *ujigami* of a shrine.
ura – during the Tokugawa period the term for fishing village.
ura-bugyō – the magistrate in charge of the fishing villages.
wakame – a seaweed (*Undaria pinnatifida*).
yakuin – a leader, a person in charge.
Yorosu-ami – a net-group based on Sakai's *sake* brewery.
Yorosuya – the merchant house operated by Sakai.

BIBLIOGRAPHY

Documents

Akaishi Documents: Kept at Fukuoka Cultural Center, Fukuoka City. Parts have been published by Komonjo o yomukai (1976) and Kaneuchi (1978). (See below.)

Isozakigū Documents: Kept at Isozaki Shrine in Shingū. Copies are kept at Fukuoka Cultural Center. Parts have been published by Komonjo o yomukai and Kaneuchi (see below).

Kaneuchi Documents: Copies are kept at Fukuoka Cultural Center, and parts have been published by Kitamura (1976) and Kaneuchi (see below).

Ketsugi-roku ōaza Shingū: Kept by the ward headman (*kuchō*) in Shingū-ku.

Sainen-ji Documents: Kept at the temple Sainen-ji, Shingū.

Yamazaki Documents: Kept at Kyūshū University, Fukuoka City. Parts have been published by Fukuoka-ken 1963 (see below).

Yayama Documents: Copies are kept at Fukuoka Cultural Center.

Books, Periodicals & Articles

Abe Kōbō: *The Woman in the Dunes* (trans. by Saunders). Tōkyō: Charles E.Tuttle Co., 1967.

Agency of Cultural affairs: *Japanese Religion.* Tōkyō: Kodansha International Ltd., 1972.

Andersen, Raoul R.: Hunt and Deceive: Information Management in Newfoundland Deep-Sea Trawler Fishing. In Andersen, R. and C. Wadel (eds) *North Atlantic Fishermen,* 1972.

Andersen, Raoul R. and Cato Wadel (eds): *North Atlantic Fishermen.* St. Johns: Memorial University of Newfoundland, Newfoundland Social and Economic Papers No. 5, 1972.

Bailey, F.G.: *Stratagems and Spoils.* Oxford: Basil Blackwell, 1969.

Barth, Fredrik: *Models of Social Organization.* London: Royal Anthropological Institute of Great Britain & Ireland, Occasional Paper No. 23, 1966.

Beardsley, Richard K.: The household in the status system of Japanese villages. In *Occasional Papers* 1:62-74. Ann Arbor: University of Michigan, Center for Japanese Studies, 1951.

Beardsley, Richard K., John W. Hall and Robert E. Ward: *Village Japan.* Chicago: University of Chicago Press, 1969.

Befu Harumi: Patrilineal Descent and Personal Kindred in Japan. *American Anthropologist* 65:1328-1341, 1963.

Bellah, Robert N.: *Tokugawa Religion – The Values of Pre-Industrial Japan.* Boston: Beacon Press, 1970.

Bennett, John W. and Ishino Iwao: *Paternalism in the Japanese Economy.* Minneapolis: University of Minnesota Press, 1963.

Bernier, Bernhard: *The Popular Religion of a Japanese Village and Its Transformation.* Doctoral dissertation, Cornell University, 1970.

Breaking the Cosmic Circle: Religion in a Japanese Village. Ithaca: Cornell University, East Asia Papers No. 5, 1975.

Blacker, Carmen: *The Catalpa Bow: A Study of Shamanistic Practices in Japan.* London: George Allen & Unwin, 1975.

Borgstrom, Georg; *Japan's World Success in Fishing.* London: Fishing News, 1964.

Brameld, Theodore: *Japan: Culture, Education and Change in Two Communities.* New York: Holt, Rinehart and Winston, 1968.

Brown, Keith: Dōzoku and the Ideology of Descent in Rural Japan. *American Anthropologist* 68:1129-1151, 1966.

Casteel, R.W. and G.I. Quimby (eds): *Maritime Adaptation of the Pacific.* The Hague: Mouton Publishers, 1975.

Chang, Kenne H-K.: The Inkyo System in Southwestern Japan: Its Functional Utility in the Household Setting. *Ethnology* 9 (4):342-357, 1970.

Institutional Changes and Development of the Fishing Industry in a Japanese Island Community. *Human Organization* 30 (2):158-169, 1971.

Chikuhō suisan kumiai: *Chikuhō enkakushi* (Historical Records of Chikuhō). Fukuoka City, 1917.

Cornell, John B.: Local group stability in the Japanese community. *Human Organization* 22:113-125, 1963.

Dōzoku: an example of evolution and transition in Japanese village society. *Comparative Studies in Society and History* 6:449-480, 1964.

Crawcour, Sidney and Yamamura Kozo: The Tokugawa Monetary System: 1787-1868. *Economic Development and Cultural Change* 18 (4):489-518, 1970.

Curtis, Gerald L.: *Election Campaigning Japanese Style*. New York and London: Columbia University Press, 1971.

Davis, K. and W.E. Moore: Some Principles of Stratification. *American Sociological Review*. Vol.10, 1945.

Dorson, Richard M. (ed): *Studies in Japanese Folklore*. Bloomington: Indiana University Press, 1963.

Ekkenkai: *Ekken zenshū* (The Collected Works of Kaibara Ekken), Vol. 4 *Chikuzen kuni zoku fudoki* (Description of Chikuzen fief). Tōkyō: Kokushokankokai, 1973.

Embree, John F.: *Suye Mura: A Japanese Village*. Chicago: University of Chicago Press, 1939.

Epstein, Scarlett: Productive Efficiency and Customary Systems of Reward in Rural South India. In Firth, R. (ed) *Themes in Economic Anthropology*, ASA Monograph 6, London: Tavistock Publications, 1967.

Firth, Raymond: *Malay Fishermen: Their Peasant Economy*. New York: W. W. Norton & Company Inc., 1966.

Freeman, J. D.: On the concept of kindred. *Journal of Royal Anthropological Institute* 91:192-220, 1961.

Fukuoka-ken: *Fukuoka-ken gyogyōshi: Chikuzen kuni* (Description of fisheries in Fukuoka Prefecture: Chikuzen fief). Unpublished manuscript kept at Fukuoka Cultural Center, 1878.

Fukuoka-ken tōkeisho (Fukuoka Prefecture Statistics). Fukuoka City, 1902, 1905, 1910, 1925.

Fukuoka-ken tōkei nenkan (Fukuoka Prefecture Statistical Yearbook). Fukuoka City, 1955, 1960, 1965, 1970, 1975.

Fukuoka-ken shi (The History of Fukuoka Prefecture), Vol.2 No.2, Fukuoka City, 1963.

Fukutake, Tadashi: *Asian Rural Society: China, India, Japan*. Tōkyō: University of Tōkyō Press, 1967.

Japanese Rural Society. London: Cornell University Press, 1972.

Gerrard, Siri: *Arbeidsliv og Lokalsamfunn*. Unpublished dissertation, Tromsø University, 1975.

Goffman, Erving: *The Presentation of Self in Everyday Life*. London: Penguin Books, 1971.

Gunnarsson, Bo: *Japans harakiri: Offren for det ekonomiska miraklet*. Stockholm: Aldus Aktuellt, 1971.

Hall, John W.: *Japan: From Prehistory to Modern Times*. Tōkyō: Charles E. Tuttle Co., 1971.

Henderson, Dan Fenno: *Conciliation and Japanese Law*, Vol.1. Seattle: University of Washington Press, 1965.

Hirschmeier, Johannes and Yui Tsunehiko: *The Development of Japanese Business 1600-1973*. London: George Allen & Unwin, 1975.

Īto Oshiro (ed): *Fukuoka-ken shiryō sōsho* (Collected Doccuments related to Fukuoka Prefecture), Vol.9. Fukuoka City: Fukuoka kenchō shomuka shiryō hensansho, 1949.

Japan Almanac: Published by Mainichi Shinbun, Tōkyō, 1976.

Johnson, Erwin: The Stem Family and Its Extension in Present Day Japan. *American Anthropologist* 66:839-851, 1964.

Status Change in Hamlet Structure Accompanying Modernization. In Dore, R. P. (ed) *Aspects of Social Change in Modern Japan*. Princeton: Princeton University Press, 1967.

Nagura Mura: An Ethnohistorical Analysis. Ithaca: Cornell University, East Asia Papers No. 7, 1976.

Kaneuchi Seiichi: *Chikuzen-kuni urakata shiryō no uchi Shingū-ura shiryō* (Shingū-ura Documents from Documents related to Fishermen in Chikuzen fief). Shingū-machi, 1978.

Kasuya-gun: *Kasuya-gun shi* (Records on Kasuya County). Fukuoka City, 1924.

Kitamura Keiko: Kaneuchi monjo ni tuite (About the Kaneuchi Documents). In Komonjo o yomukai *Chikuzen-kuni urakata shiryō*. Fukuoka City, 1976.

Kōhō Shingū-machi: Monthly paper published by Shingū-machi, Vol.4, 1976.

Komonjo o yomukai: *Chikuzen-kuni urakata shiryō* (Documents related to Fishermen in Chikuzen fief). Fukuoka City: Fukuoka Cultural Center, 1976.

Kujukuri Research Committee: *Kujukurihama: Study of Fishing Community in Japan*. Tōkyō: Keio University (mimeo), 1958.

Kyūshū Nippō Shinbun: Newspaper published in Kitakyūshū City, Vol. 1-13, 1935.

Linton, Ralph: *A Study of Man*. New York, 1936.

Lofgren, Orvar: Resource Management and Family Firms: Swedish West Coast Fishermen. In Andersen, R. and C. Wadel (eds) *North Atlantic Fishermen*. 1972.

Mainichi Daily News: 11 July and 9 November 1975.

Masuda Kōkichi: Bride's Progress: How a *Yome* becomes a *Shūtome*. *Journal of Asian and African Studies* X, 1-2: 10-19, 1975.

Mathias, R.: The recruitment and the organization of labour in the coal mining industry of northern Kyushu during the Meiji period. *Orientaliska Studier* 24-25: 50-59, 1976.

Monthly Statistics of Japan: Published by Bureau of Statistics, Office of the Prime Minister, Tōkyō, June 1978.

Morioka Kiyomi: The Impact of Suburbanization on Shinto Belief and Behaviour. *Social Compass* XVII:37-65, 1970/1.

Moyer, Jack T.: Bounty of Kuroshio. In Beardsley, R.K. (ed) *Studies in Japanese Culture 1*. Ann Arbor: University of Michigan Press, Center for Japanese Studies, 1965.

Naitō Kanji: Inheritance Practices on a Catholic Island: Youngest-Son Inheritance (Ultimogeniture) on Kuroshima, Nagasaki Prefecture. *Social Compass* XVII:21-36, 1970/1.

Nakamura Masao: Higo Amakusajima ni okeru gyoson no seiritsu to tenkai: Funakoyaku o chūshin to site (Formation and development of fishing villages on Amakusa Islands in Higo fief, with focus on *funakoyaku*). Fukuoka City: Kyūshū University, 1961.

Nakane Chie: *Kinship and Economic Organization in Rural Japan*. London: The Athlone Press, 1967.

Japanese Society. Penguin Books, 1973.

An interpretation of the size and structure of the household in Japan over three centuries. In Laslett, P. (ed) *Household and Family in Past Time*. London: Cambridge University Press, 1974.

Nemec, Thomas F.: I Fish With My Brother: The Structure and Behaviour of Agnatic-Based Fishing Crews in a Newfoundland Irish Outport. In Andersen, R. and C. Wadel (eds) *North Atlantic Fishermen*. 1972.

Norbeck, Edward: Age-grading in Japan. *American Anthropologist* 55:373-384, 1953.

Takashima: A Japanese Fishing Community. Salt Lake City: University of Utah Press, 1954.

Little-Known Minority Groups of Japan. In de Vos, G. and H. Wagatsuma (eds) *Japan's Invisible Race: Caste in Culture and Personality*. Berkeley and Los Angeles: University of California Press, 1967.

Ōtō Tokihiko: The Taboos of Fishermen. In Dorson, R. M. (ed) *Studies in Japanese Folklore*. 1963.

Papinot, E.: *Historical and Geographical Dictionary of Japan*. Tōkyō: Charles E. Tuttle Co., 1972 (first published 1910).

Pyle, Kenneth B.: The Technology of Japanese Nationalism: The Local Improvement Movement, 1900-1918. *Journal of Asian Studies* 33 (1):51-65, 1973.

Sahlins, Marshall, D.: *Stone Age Economics*. London: Tavistock Publications, 1974.

Sansom, George B.: *Japan: A Short Cultural History*. Revised edition, London: Barrie & Jenkins Ltd., 1946.

A History of Japan to 1334. Tōkyō: Charles E. Tuttle Co., 1974a.

A History of Japan, 1334-1615. Tōkyō: Charles E. Tuttle Co., 1974b.

Schepers, Hansjulius: *Japans Seefischerei: Eine wirtschaftsgeographische Zusammenfassung*. Breslau: Ferdinand Hirt, 1935.

Segawa Kiyoko: Menstrual Taboos Imposed upon Women. In Dorson, R. M. (ed) *Studies in Japanese Folklore*. 1963.

Seiko Teruo: *Gyogyō no rekishi* (The History of Fishing). Tōkyō, 1966.
Sheldon, Charles D.: *The Rise of the Merchant Class in Tokugawa Japan, 1600-1868.* New York: J. J. Augustin, 1958.
Shibusawa, K. (ed): *Japanese Life and Culture in the Meiji Era.* (trans. by Terry.) Tōkyō: Centenary Cultural Council Series, 1958.
Smith, Robert J.: Small families, small households, and residential instability: town and city in "pre-modern" Japan. In Laslett, P. (ed) *Household and Family in Past Time.* London: Cambridge University Press, 1974a.
Ancestor Worship in Contemporary Japan. Stanford: Stanford University Press, 1974b.
Smith, Thomas C.: The Japanese Village in the Seventeenth Century. *The Journal of Economic History* XII:1-20, 1952.
The Agrarian Origin of Modern Japan. Stanford: Stanford University Press, 1959.
Nakahara: Family Farming and Population in a Japanese Village, 1717-1830. Stanford: Stanford University Press, 1977.
Smith, W. Eugene and Aileen M. Smith: *Minamata.* London: Chatto & Windus Ltd., 1975.
Southall Aidan: Ideology and Group Composition in Madagascar. *American Anthropologist* 73:144-164, 1971.
Statistical Handbook of Japan: Published by Bureau of Statistics, Office of the Prime Minister, Tōkyō 1975.
Suisan Tsūshinsha: *Suisan meikan* (Directory of Fisheries). Tōkyō, 1939.
Taira Koji: *Economic Development and the Labor Market in Japan.* New York: Columbia University Press, 1970.
Tanizaki Junichirō: *The Marioka Sisters* (trans. by Seidensticker). Tōkyō: Charles E. Tuttle Co., 1969.
Tsukahira, T. G.: *Feudal Control in Tokugawa Japan.* Cambridge: Harvard University Press, 1966.
Tsukushi Yutaka: *Nihon no minzoku: Vol.40 Fukuoka* (Japanese Folklore: Vol.40 Fukuoka). Tōkyō: Daiichi hōki shuppan kabushiki-gaisha, 1974.
Vogel, Ezra: *Japan's New Middle Class.* Berkeley: University of California Press, 1963.
Wadel, Cato: Capitalization and Ownership: The Persistence of Fishermen-Ownership in the Norwegian Herring Fishery. In Andersen, R. and C. Wadel (eds) *North Atlantic Fishermen,* 1972.
Wigmore, John Henry: *Law and Justice in Tokugawa Japan,* Part I. Tōkyō: University of Tōkyō Press, 1969.
Wolf, Eric: Kinship, Friendship, and Patron-Client Relations in Complex Societies. In Banton, M. (ed) *The Social Anthropology of Complex Societies.* ASA Monograph 4, London: Tavistock Publications, 1966.
Yamaoka Eiichi: *Gyoson shakaigaku no kenkyū* (Studies on the Sociology of Fishing Villages). Tōkyō: Daimeidō, 1974.
Yanagita Kunio: *About Our Ancestors: The Japanese Family System.* Tōkyō: Japan Society for the Promotion of Science, 1970.
Yazaki Takeo: *Social Change and the City in Japan.* Tōkyō: Japan Publications Inc., 1968.